P9-CEZ-039

Musings

Musings

The Musical Worlds of Gunther Schuller

GUNTHER SCHULLER

New York Oxford
OXFORD UNIVERSITY PRESS
1986

OXFORD UNIVERSITY PRESS

Oxford New York Toronto
Delhi Bombay Calcutta Madras Karachi
Petaling Jaya Singapore Hong Kong Tokyo
Nairobi Dar es Salaam Cape Town
Melbourne Auckland

and associated companies in
Beirut Berlin Ibadan Nicosia

Published by Oxford University Press, Inc.,
200 Madison Avenue, New York, New York 10016

LIBRARY OF CONGRESS CATALOGING-IN-PUBLICATION DATA
Schuller, Gunther.
Musings : the musical worlds of Gunther Schuller.
1. Music—Addresses, essays, lectures. I. Title.
ML60.S392 1986 780 85-15464
ISBN 0-19-503745-6

Printing (last digit): 9 8 7 6 5 4 3 2 1
Printed in the United States of America

For Bruce Creditor

For permission to reprint articles from the books, publications and records indicated, grateful acknowledgment is made to the following:

Saturday Review magazine for "The Future of Form in Jazz" © 1957 Saturday Review magazine and "Third Stream" ["Third Stream Redefined"] © 1961 Saturday Review magazine. Used by permission.

W. W. Norton & Company for "James Reese Europe" reprinted from *Dictionary of American Negro Biography,* Edited by Rayford W. Logan and Michael R. Winston. © 1982 by Rayford W. Logan and Michael R. Winston. Used by permission.

Golden Crest Records for "Happy Feet: A Tribute to Paul Whiteman," liner notes from record CRS 31043. Used by permission.

High Fidelity for "Ellington in the Pantheon" (Nevember 1974), "The Case for Ellington's Music as Living Repertory (Are the Recordings Enough?)" (November 1974), and "the State of American Orchestras" ["The Trouble with Orchestras"] (June 1980). All rights reserved. Used by permission.

Smithsonian Collection of Recordings, a division of the Smithsonian Institution Press for "Ellington vis-à-vis the Swing Era" first published as liner essay for *Duke Ellington 1938,* Smithsonian Collection R003. Used by permission.

The Jazz Review for "Cecil Taylor: Two Early Recordings" (January 1959) and "Sonny Rollins and the Challenge of Thematic Improvisation" (November 1958). Used by permission.

MJQ Music for "Ornette Coleman's Compositions," reprinted from "A Collection of Compositions by Ornette Coleman" © 1961, MJQ Music, Inc. Used by permission.

Fantasy, Inc. for "Lee Konitz," liner notes from MSP 9013. Used by permission.

New World Records for "The Avant-Garde and Third Stream" which originally appeared in liner notes to *Mirage: Avant-Garde and Third Stream Jazz,* New World Records NW 216. © 1977 Recorded Anthology of American Music, Inc. All rights reserved. Used by permission.

University of Oklahoma Press for "Composing for Orchestra" from *The Orchestral Composer's Point of View: Essays on Twentieth-Century Music by Those Who Wrote It,* edited by Robert Stephan Hines. © 1970 the University of Oklahoma Press. Used by permission.

Perspectives of New Music for "American Performance and New Music" (Volume 1, number 2, Spring 1963). © 1963 Perspectives of New Music, Inc. Used by permission.

High Fidelity for "Ellington in the Pantheon" (November 1974), "The Case for 1974), and "The State of American Orchestras" ["The Trouble with Orchestras"] McGraw-Hill Book Company for "Conducting Revisited" from *The Conductor's Art,* edited by Carl Bamberger. © 1965 McGraw-Hill, Inc. Used by permission.

Opera News for "The Future of Opera" (May 1967). Reprinted courtesy of Opera News, a publication of the Metropolitan Opera Guild. Used by permission.

Keynote Magazine for "The State of Our Art" (May/June 1982). Used by permission.

Philosophical Library for "Form and Aesthetics in Twentieth-Century Music" from *The Aesthetic Dimension of Science (1980 Nobel Conference).* Used by permission.

Foreword

FOR THOSE OF US who well remember Gunther Schuller as a young man with a horn, that he is now sixty evokes the customary shock of chronological disbelief; for those who know the name of Gunther Schuller only by its multiform musical presences and his manifold accomplishments, it will come as a comparable surprise to learn that he is only sixty.

When, in 1967, Gunther ascended, or descended to the Presidency of the New England Conservatory (most likely and characteristically, he travelled in both directions simultaneously) it could not have been foreseen that his tenure in that position would be almost exactly equal to that of Liszt's at Weimar, but it could have been anticipated that, during that tenure, he would surpass Liszt's record for shepherding works to their first performance, rare performance, or appropriate performance. For when, in 1970, Gunther received the Alice M. Ditson conducting award for his "unselfish championship of fellow composers," those words seemed too modest and reserved to a serious fault, for they reflected so little of the degree and extent of the championship and of the range of the musical dispositions of the "fellow composers."

Gunther's championship seasons had begun at least a quarter of a century earlier, when the young horn player somehow prevailed upon four of his associates in the Metropolitan Opera Orchestra, who probably never had been aware of the very existence of the Schoenberg Wind Quintet, to learn that demanding, refractory composition and to record it for the first time, some twenty-five years after its composition. This ability to induce reluctant, even apprehensive, performers to apply themselves sedulously, even eventually enthusiastically, was a crucial component of the versatile virtuosity which Gunther then brought to that unprecedented series of concerts "Twentieth Century Innovations," which, for three seasons in the early sixties, presented expert performances of a range of music which reflected the total of that singular pluralism which was and is our legacy and condition. Gunther selected the works and the performers, rehearsed them, and conducted all of the compositions which required a conductor. The concerts were, in spirit, effect, and duration, the legitimate heir of the Viennese Society for Private Performance, particularly as expressed in Schoenberg's words in its statute, that "the success which the composer is to have here is that which should be of the greatest importance for him: to be in the position to make himself understood."

At that time Gunther extended his activity to oral performance in a series of over 150 broadcasts, in which he presented, by commentary and recording, a survey of twentieth-century music from its roots to its latest bloomers. It was not only a labor of love, but a labor of urgency, for he felt that professional survival was dependent upon reaching out and securing support from a mainly verbally oriented society by providing sensible words about the sounds of the music.

A decade later a brilliantly produced series, similarly fusing talk and—now—"live" performance, for Public Televison, went too unheard and unseen in many cities, including New York, where its scheduled appearances were replaced by the tediously detailed, continuous presentation of a Congressional investigation of a petty political peccadillo, providing yet again evidence of the status of serious contemporary music in the world of media.

If it is the Gunther of the "third stream" and its environs who is best known to the general public, that would have seemed the most unlikely Gunther when he began his career, as a teen-ager, as principal horn of the Cincinnati orchestra. But it was there that he did commit a glittering, Gould-Kostelanetz-like transcription of "Night and Day" for the orchestra, and—having savored the previously forbidden fruit of "popular music"—the young musician, who had been cocooned in a pristinely classical environment by family and education, proceeded to play, transcribe, write about, and incorporate into his own music jazz in *all* its fragmentations. Later, at the New England Conservatory, he formed and directed the celebrated Ragtime Ensemble, the Jazz Repertory Orchestra, the Country Fiddle Band, and then he wrote *Early Jazz:* analysis, criticism and history unrivalled in depth, scope, and scrupulousness.

In these collected articles much of Gunther is present, and Gunther the composer always is present, for all that he has done is shaped by the awarenesses of the composer. In so explicitly compositionally centered an article as "Composing for Orchestra," one can observe the subtle interplay and interaction between Gunther's thinking in music and thinking about music. If this is one of the more "technical" (I would prefer to say "musical") articles, it is—at the same time—one of the personally most revelatory.

Gunther the conductor appears here in various guises, and I hope only that the reader can infer something of the singular capacities of a conductor who can achieve maximal performances under minimal conditions, by knowing the score and knowing the performers, be they members of a celebrated "major" orchestra or of a student orchestra. His rehearsals, then, are carefully planned musical and psychological journeys and lessons. He does not "take it from the top," and—at reduced tempo—plow and struggle through a work, thus providing the players with evaluative justification for their hostility. Rather, he begins with selected non-consecutive passages, often the most difficult, thereby reassuring the orchestra of its ability eventually to perform the work, while placing the players in their proper roles as professional performers, rather than as amateur critics. And as for the apparently more contentious articles on the state of the "professional" orchestra, at this chronological remove it can be understood why only at the time of their appearance were they regarded only as timely; the conductors may go, the conditions remain.

Only Gunther's two most recent undertakings are not explicitly documented here: music publishing and recording. But these least visible enterprises are among his most vital and influential advocacies. When Brahms asserted that he would not judge a work until he had seen a score, scores were being published. Today, for a terrifying variety of economic reasons, music publishing is a disappearing act of faith; the number of works published is in inverse ratio to the number produced. The already extensive catalogues of Margun and Gunmar embody the energy and discriminating catholicity that are the earmarks of Gunther's composing, conducting, teaching, and—of course—writing. And since what is published and recorded is what is heard and known, it necessarily ultimately determines what is composed. So those historians of music who wait for time to tell them will not have long to wait to tell the extent to which Gunther has shaped the history of music in our time.

Milton Babbitt

Contents

I. JAZZ AND THE THIRD STREAM

II. MUSIC PERFORMANCE AND CONTEMPORARY MUSIC

III. MUSIC AESTHETICS AND EDUCATION

Jazz and the Third Stream

1

Jazz

A short history of jazz written in 1976 for a prospective encyclopedia—which never materialized. Although limited by space restrictions, the article still manages to cover the high points and "major stations" in the eight-decade development of our one and only indigenous American music.

JAZZ. Jazz is a style of music first developed in the southern United States representing a confluence of West African and European musical-social traditions. Although jazz is often notated (in compositions or arrangements), its quintessential means of expression is improvisation, an inevitable consequence of its origins as an Afro-American folk music. The term "jazz" (at times spelled "jass"), originally a common vulgarism used in American red-light districts, began to be applied around 1916-17 to a music already in existence in a fairly integral form for nearly two decades. This music, essentially improvised and played by black musicians who were prohibited by social and racial barriers from acquiring a formal musical education, was in turn a blend of earlier folk and popular elements, foremost among these the "blues" and "ragtime."

Ragtime was in itself a product of black and white acculturation, developed in the American Midwest in the 1890s—its foremost composer was Scott Joplin—by amalgamating European forms (march, quadrille), harmony and symmetric rhythms or meters with African asymmetrical rhythmic patterns. Originally a fully notated and non-improvised piano music, ragtime gradually spread to other instruments, intersecting at this point with another venerable American musical tradition, that of the marching or brass band. This preeminent late nineteenth-century American musical institution was widespread geographically and indigenous to many ethnic and immigrant cultures. Indeed, it was the influence of French-style military bands in New

Orleans and the availability of band instruments at war-surplus prices after the Civil and the Spanish-American wars that precipitated the emergence of the black marching band tradition in the southern United States. The formal, stylistic proximity of marches to ragtime undoubtedly led early brass instrumentalists like Buddy Bolden, Bunk Johnson, and Kid Ory to adapt ragtime pieces to their own musical needs.

But rhythmically "ragging" melodies and themes was only one step removed from loosening them up even further through improvisation and melodic embellishment. Thus, many of the earliest jazz musicians were essentially ragtime players, or, to put it more precisely, musicians who were transitional in the progress from a relatively rigid, notated, non-improvised music (ragtime) to a looser, more spontaneously inventive performance style (jazz). This process succinctly delineates the fundamental difference between ragtime (and other notated musics, such as "classical" music) and jazz: in the former, it is the composition per se, the "what," that is crucial; in the latter, it is the performance expression, the "how," that counts.

The origins of the "blues," the other distinct pre-jazz source, are lost in obscurity. Our best guess is that the blues represents a coalescence some time before the turn of the century of field hollers, work songs, ring shouts (all of African heritage), and the spiritual, in itself a crossing of African melody (primarily pentatonic) with Anglo-Saxon hymn harmonies. Bard-like itinerant country blues singers (many of them blind) traveled the width and breadth of the southwestern United States for decades before the blues, in the wake of southern industrialization, eventually settled down in urban centers and became standardized as a specific three-part stanza form and a closely related three-step harmonic progression (I, IV, V).

Perhaps no major figure better exemplifies the cross-fertilization of the blues and ragtime than Ferdinand "Jelly Roll" Morton. As a highly original composer (creator) and an indefatigable performer (re-creator), he embodied already early in his career the synthesis of ragtime—with its instrumental and formal heritage—and the blues—with its essentially vocal and improvisatory traditions—the two emerging as a music called "jazz" by the time of World War I.

The elements that make jazz a distinct musical language are manifold and largely of African descent. (This accounts for the fact that to this day all major innovations in jazz are attributable to Afro-Americans.) The most easily identifiable and distinguishing element of jazz is the rhythm. By superimposing the highly complex polyrhythmic and polymetric structures of African music on the comparatively simply binary rhythms and meters of European music, the American Negro was creating a unique and unprecedented musical symbiosis. The inherent conflict between rhythmic symmetry and asymmetry, the constant tension between a steady metronomic underlying beat and the unnotable, infinitely subtle permutations (especially when improvised) of polyrhythms, the equilibrium maintained between strict controlled tempo and relaxed rhythmic spontaneity—these are the essential energizing antipodes of jazz.

The result was called "syncopated music" in the 1920s and "swing" in the 1930s. But by whatever name it may be called, jazz inflection is based on (1) the simultaneous feeling of *both* antipodal rhythmic levels and (2) maintaining a perfect equilibrium between the horizontal and vertical relationships of musical sounds. When the horizontal and vertical aspects of notes, placed in a time continuum, are heard as but two parts of the same impulse, then we have "swing," the rhythmic language of jazz. To put it another way, swing results from a seemingly contradictory ambivalence—perceived and applied simultaneously—of tension-laden rhythmic *control* and the utmost *spontaneity*. Rhythmic precision per se does not constitute swing; it is when rhythmic precision is expressed in terms of natural and relaxed rhythmic impulse that the essence of swing is likely to be achieved.

But the African heritage reveals itself in more than just the rhythmic aspects of jazz. Taking melody and harmony together, since they are but different expressions of the same element, pitch—again the horizontal and the vertical—we note that the African's pentatonic and heptatonic scales were easily accommodated within the European framework of diatonic harmony. To the extent that we can speak of African harmony in the Western sense—diodic organum would be a more accurate term—this, too, was readily assimilated into the diatonic system. The most interesting progeny of this cultural assimilation is the blues scale, with its "blue notes" on the third and seventh degrees (square notes):

EXAMPLE 1

To qualify as a "blue note," the third and seventh can be played (or sung) with a variable intonation ranging all the way from the minor to the major, coinciding very nicely, of course, with the European concept of minor and major tonalities. And when, with increasing sophistication, jazz absorbed European chromatic harmony, leading eventually to simple forms of bitonality, it was still on African home territory, for the simultaneous sounding of *both* the (blue) minor third and the major third (as in Ex. 2, the two squared notes) constitutes one of the more fundamental of bitonal chords (a combination of E-flat and C major).

EXAMPLE 2

In the realm of timbre, too, jazz's African heritage survives. The African open tone and "natural" quality in speech, song and playing finds its American echoes in the earthy, sonorous voices of countless blues singers, the individuality and personal inflection of hundreds of jazz artists from Armstrong to Ornette Coleman. A byproduct of this unique sense of timbral individuality is the invention by jazz brass players of an astonishing variety of mutes and other timbre-changing devices, including the growl-and-plunger techniques pioneered by Bubber Miley and Cootie Williams, and beyond even those, the virtual "talking" style of "Tricky Sam" Nanton. That these timbral characteristics have survived, despite the fact that jazz developed almost entirely on instruments of European origin, is perhaps one of the more remarkable achievements of jazz.

The basic forms of jazz reveal strong African antecedents, particularly in the "call and response" pattern, in the "chorus" concept and in the "riff" principle. The first of these permeates almost all African music, usually taking the form of a chorus responding to a leader or "liner." In Afro-American music, this device can be heard not only in jazz arrangements from Fletcher Henderson to Count Basie, but in gospel music and the church's minister-congregation dialogues, and even in the basic blues structure with its two "calls" and one "response."

The "chorus" concept—virtually unknown in European music—taken directly from the "master patterns" of tribal cult and recreational dances, is perhaps the most fundamental structural principle to jazz. It is as the sonata form is to diatonic music.

The riff, a kind of repetition or ostinato device, corresponds exactly to the repetitive structuring of African work and play songs, and to the secondary drum patterns in African ensemble music.

Thus jazz, in all its essential early forms, represents an exact musical analogy to the Afro-American's preservation of most of his social, cultural, and moral values (while assimilating within a white societal structure).

The "classic" period of jazz (roughly from 1905 to 1925) is divisible into three styles, the most important of which is associated with the geographic cradle of jazz, New Orleans, the other two being Chicago-style and Dixieland. During this period jazz evolved from its earlier "folk music" origins into an entertainment music. The relatively relaxed racial attitudes and social conditions of liberal New Orleans made it possible for blacks, particularly the Creoles, to integrate into the lively social and entertainment life of the city. In the aristocratic salons, on the river boats, in the bars and bordellos of the Vieux Carré, and in the funeral processions and street parades, music was in constant demand.

By the mid-1890s a standard New Orleans ensemble had evolved, comprising the "front line" of cornet, clarinet, trombone, and a rhythm section consisting of bass (at first tuba, later string bass), banjo or guitar (the piano was a late arrival because it was a relatively cumbersome instrument to carry in street parades), and drums, originally a collection of simple percussion instruments—usually bass drum, snare drum, and cymbal—called a "trap-set" and actually invented in New Orleans by some now-long-forgotten, self-taught drummer.

For musicians lacking in formal musical education and therefore unable to read or write music, recourse to improvisation was the only way to musical self-expression. The spontaneity of instant creation was the driving force behind New Orleans jazz, to which was added the concept of "collective" improvisation. The essentially homophonic dance and march music of the period was transformed into polyphonic, improvised jazz. Although the three melodic strands of the "front line" would occasionally criss-cross, and though the music was apt to be full of unpredictable rhythmic cross-accents, collective improvisation worked because (1) there were only three melodic instruments, each restricted more or less to certain registers and functions (the cornet most frequently outlining the melody, the trombone supporting with counterstatements in the tenor register, and the clarinet embellishing the cornet's line in flowing descant fashion); (2) the improvisations were based on relatively simple harmonic progressions. The epitome of this style was achieved in the playing of *King Oliver's Creole Jazz Band* (recorded in 1923), Louis Armstrong's *Hot Five* and *Hot Seven* recordings (1925-27) and, in some respects in its most sophisticated form, by Morton's *Red Hot Peppers* recordings of 1926-28. In these recordings, jazz made the dramatic step from an entertainment music to an art form.

Ironically, the very skills required to master collective improvisation and simple "head arrangements" spelled the doom of the pure New Orleans style. The virtuosity, soloistic prowess, and sheer individuality of a Louis Armstrong or a Sidney Bechet prophetically carried the seeds of a new direction in jazz: the solo concept, and with it the need to arrange or compose a framework in which the soloist(s) could operate.

The New Orleans style, except for its revival in the late 1930s and early 40s, survived in two lesser derivatives, both primarily of white manufacture. When the red-light district of New Orleans, known as Storyville, was closed down by the U.S. Navy in October 1917, jazz moved out and "up the river" to Chicago. There, young white musicians began to emulate their black idols (Oliver, Armstrong, Dodds, Noone, etc.), not without, however, a loss in the original stylistic purity. The most famous group of Chicago-style musicians, the so-called Austin High School Gang—comprising cornetist Jimmy McPartland, clarinetist Frank Teschemacher, saxophonist Bud Freeman, drummer Gene Krupa (Bix Beiderbecke, cornet, and Benny Goodman, clarinet, also must be counted as charter members of the Chicagoans)—added the saxophone to the collective ensemble, thus in part sacrificing the textural clarity of the traditional New Orleans "front line." As men of lesser talent than their black mentors, they lost much of the warm, relaxed swing of New Orleans, substituting for it a hard, nervous "jazz age" temper that clearly symbolized the difference between Chicago (in the North) and New Orleans (in the South). Without the inner motivation of the black New Orleans innovators, the Chicagoans succumbed to the subtle (or often not so subtle) pressures of the commercial marketplace shaped by record sales, the economic success of dance and popular orchestras, and the encroachment of Tin Pan Alley: in short, jazz as a big business organized and controlled by managers and, in the "prohibition era," subsidized by gangsters.

The term "Dixieland" first came into vogue with the formation of the Original Dixieland Jazz Band, the first jazz group to record (in 1917) and an overnight sensational success in New York during the same year. A tour in Europe in 1918 spread the word "jazz" to the mother continent and thence to the rest of the world. Again, it is ironic—and symbolic of the struggle for jazz as an Afro-American music to survive in a white marketplace—that it was the crude, musically primitive vulgarisms of the ODJB which trumpeted jazz, at least in name, to the world at large.

In later years, the term "Dixieland jazz" has come to take on a broader meaning, from a specific association with fine musicians like Muggsy Spanier (cornet), Jack Teagarden (trombone), Eddie Condon (guitar), and Pee Wee Russell (clarinet), to a very broad catch-all reference to any form of traditional jazz, emulating or reviving the old New Orleans style, ranging from the amateur to the frankly commercial bands dressed in bowler hats and Bermuda shorts.

By the mid-20s, Armstrong, the first major soloist of jazz, had revolutionized the music. It took only a few years before the demise of New Orleans jazz was complete, aided and abetted by a new generation of players, by postwar industrial and population shifts, by the rapid, massive geographical dissemination through records, radio, and film of new musical styles, and finally the great Crash of 1929 and the ensuing Depression.

But in the halcyon days of Chicago in the late twenties, the jazz mainstream had already bifurcated: one to New York, the other to Kansas City. The former was the financial and commercial capital of the country; the latter became the newest manifestation of ganster-ridden city political machines as first pioneered in Chicago. Both cities drew the best musicians like magnets.

In New York, Fletcher Henderson and his chief arranger, conservatory-trained Don Redman, had been evolving an expanded orchestral style since 1924, far removed from New Orleans polyphony. Indeed, the very size of the numerically larger orchestras mitigated against collective improvisation. As a result, a new breed of musician who could read at least simple arrangements was developing. And those, like saxophonist Coleman Hawkins and Armstrong, who could combine reading with great solo improvisations, comprised the new artistic leadership. Through half a dozen years the Henderson orchestra evolved a big band style that eventually became the arranging formula for an entire decade, the Swing Era. As the orchestras grew in size to a standard instrumentation of fourteen (3 trumpets, 3 trombones, 4 saxophones, and 4 rhythm), it was the arranger's task to create the personal style of the band. Of the several devices that Redman perfected, one was the fragmentation of the band into timbral choirs—a particular Redman favorite was the clarinet trio—and utilizing these choirs in terms of the old "call and response" pattern (as in the Henderson band's various recordings of "King Porter Stomp"). By the end of the Swing Era, the Henderson formula had stifled creativity, not only because it left less and less room for the soloist but because of the growing rigidity in the incessantly homophonic block writing.

Relief was to come from the other offshoot of the jazz mainstream, the Southwest and its musical capital of Kansas City. Here Bennie Moten had been holding forth since 1918, originally leading a small ragtime trio which had been expanded to six players by the time Moten began recording in 1923. They were by then also playing mostly blues, with a heavy, stomping 4/4 beat, quite different from the lingering two-beat rhythms of New Orleans, Chicago, and New York. The blues, of course, were deeply rooted in the whole Southwest territory of Texas, Oklahoma, and Arkansas, where for generations itinerant blues singers had criss-crossed the rural areas. In the simple, uncluttered harmonic frame of the blues, the soloists could expand their improvisational skills; a heavy, rocking, blues-drenched sense of swing could be developed; and above all, the riff, that aforementioned device of African heritage, could be elaborated with increasing sophistication until it too became a fundamental jazz-orchestral technique of the Swing Era.

Moten was not without competitors, even in his home territory. Foremost among them was Alphonse Trent (originally from Arkansas, but an immense success in Dallas, Texas). Also headquartered in Texas were Terrance Holder and Troy Floyd, while Jesse Stone roamed the Missouri and Kansas areas, and Walter Page's Blue Devils occupied Oklahoma. Then there were the eastern bands of Henderson and Ellington, and the Detroit-based McKinney's Cotton Pickers to worry about. When the competition became too tough, Moten, who was a better businessman than pianist, began buying up arrangements, particularly those of Horace Henderson (Fletcher's brother) and of one of Henderson's ace arrangers, saxophonist Benny Carter. When that didn't suffice, Moten tried to buy out his spunkiest competitor,

Walter Page's Blue Devils, who had tendered Moten a disastrous beating in a gigantic "cutting contest" in 1928. When in turn that failed, Moten raided the Page band one by one, enticing the players away with higher salaries. Alto saxophonist Buster Smith, trombonist-guitarist arranger Eddie Durham, trumpeter Hot Lips Page, and blues singer Jimmy Rushing soon left the Blue Devils, as did pianist Count Basie, eventually to be followed by the Blue Devils' leader, Walter Page. By the time Moten recorded in Camden, New Jersey, for Victor in December 1932, he had built a remarkable orchestra by adding further major talents in the persons of clarinetist-arranger Eddie Barefield and tenor saxophonist Ben Webster. These sidemen not only had a great creative talent, but they represented a "modern," new breed of post-New Orleans musician; and above all, coming from the blues-saturated Southwest, their playing exhibited a wholly new sense of swing which once and for all stamped out any lingering vestiges of Moten's ragtime beginnings.

All this can be heard on the best of Moten's 1932 sides, particularly the fast pieces: "Toby," "Blue Room," "Prince of Wales," and "Lafayette," astonishing displays of technical and ensemble virtuosity, combined with enormous rhythmic drive and soloistic inventiveness.

Unfortunately, the eastern-based recording companies neglected the southwestern territory bands. Some were never recorded at all, others only sparsely. But Moten's 1932 recordings, along with earlier ones by Trent ("After You've Gone," "I Found a New Baby"), the Blue Devils' "Squabblin' " and "Blue Devil Blues," Troy Floyd's "Dreamland," and Stone's "Starvation Blues" represent the best of that particular period and style. This tradition was to be carried on, after Moten's death in 1935, by Count Basie, through whom the Swing Era was to gain a much-needed infusion, and the word "swing" was to take on a new meaning. And that same Kansas City lineage led directly to Charlie Parker and the Bop Era of the 1940s.

Virtually outside and independent of the two main orchestral styles (Henderson/New York, Moten/Kansas City), Duke Ellington created a wholly original compositional style and a new conception of the old New Orleans idea of collective creativity. After initial years of struggle in New York in the early 1920s, two events turned out in retrospect to be crucial turning points in the development of the young Ellington band. One was the hiring of trumpeter Bubber Miley, who was not only the first major soloist Ellington acquired and a creative personality even more substantial at first than Ellington himself but was also the perfector of the "growl and plunger" style of brass playing which formed the basis for Ellington's "jungle style," a trademark his orchestra has maintained even unto today. To Miley's influence we owe the two great early Ellington masterpieces, "East St. Louis Toodle-Oo" and "Black and Tan Fantasy."

The other event was the hiring of the Ellington orchestra in 1927 by the Cotton Club in Harlem for an unprecedented five-year engagement. Its importance lay in the fact that the band could enjoy financial stability and therefore, not only relative permanence of personnel, but the opportunity to grow musically, to experiment and test; in short, to enjoy a prolonged "workshop" period. It was during these years that Ellington's own compositional talent blossomed, inspired at first by his players—he has always referred to his orchestra as his true instrument—but as well by the necessity of inventing a constant series of musical numbers of all kinds, from dance routines to production numbers, from background materials for the club's exotic, junglistic tableaux to more abstract, purely compositional efforts. During this five-year workshop period, Ellington produced a staggering amount of music: some 170 recordings, comprising over a hundred compositions or arrangements.

It was during these years that Ellington and the orchestra developed, by trial and error, the basic Ellington style, which is based as much on the individual qualities of his brilliant complement of players (including Johnny Hodges, alto saxophone; Barney Bigard, clarinet; Harry Carney, baritone saxophone; trumpeters Cootie Williams, who replaced Miley, and Rex Stewart; trombonists "Tricky Sam" Nanton and Lawrence Brown; drummer Sonny Greer) as on Ellington's own formidable compositional and organizational gifts. Thus, the concept of collective, collaborative creativity—a hallmark of New Orleans jazz—was broadened and brought by Ellington and his musicians to a new zenith.

Ellington's productivity over the last fifty years is so prodigious that an overview is virtually impossible. His uniqueness lies not only in his compositional creativity but in the Ellington orchestra's longevity of life. Both are unprecedented and unmatched in the history of jazz. These Ellington careers are marked by an endless succession of jazz masterpieces: from the 1930 "Mood Indigo" through "Daybreak Express" (1934), "Reminiscin' in Tempo" (1935), "Echoes of Harlem" (1936), "Caravan" (1937), "Blue Light" (1938) to the greatest period of all, 1940-43, which brought forth "Ko-Ko," "Jack the Bear,"

"Dusk," "Warm Valley," "Take the A-Train," "Cotton Tail," and the 50-minute orchestral suite "Black, Brown and Beige."

Sadly, however, the general American public was unaware of these latter developments in New York and Kansas City. The most popular figure in the 1930s—and perhaps even to this day—was not Basie, nor Ellington, nor Henderson, but Benny Goodman. Crowned the "King of Swing," this virtuoso clarinetist and most demanding of band leaders adopted and then perfected the Henderson-Redman arranging style, parlaying it into one of the greatest financial and popular successes jazz has ever known. In the wake of Goodman's success, literally hundreds of bands followed suit, ranging from those like Tommy Dorsey, Artie Shaw, Charlie Barnet, the Casa Loma Orchestra—all of whom retained at least a basic relationship to jazz and retained a complement of soloists to keep improvisation alive—to those who were barely more than commercial dance bands, albeit sometimes of high technical quality, like Glenn Miller and Harry James.

By the end of the 1930s, various reactions to the excesses of the Swing Era were beginning to set in. The Henderson arranging style, reduced now to a mere formula, had arrived at total stagnation. With improvisation largely suppressed and arrangements becoming increasingly empty and overblown, musicians turned elsewhere for their creative survival. Two solutions presented themselves: the "jam session" and the "combo." In a development astonishingly parallel to that in classical music shortly after the turn of the century, when composers recoiled from the overblown forms and instrumentations of the post-Romantic period, jazz musicians now retreated to a kind of chamber music approach to jazz: the five- or six-piece combo. And for those for whom commercial pressures still led to artistically stifling professional situations, there was the late-night jam session, where musicians improvised for pleasure and tested their skills in a setting of competition and camaraderie, unencumbered by managerial or musico-political constraints.

A third reaction was the "revival movement," an attempt to take jazz back to its New Orleans origins. This was sparked by a California musician, Lu Watters, who with his Yerba Buena Jazz Band began to perform, in verbatim copies, the old Armstrong, Oliver, and Morton standards. Before the movement ran its course fifteen years later, Bunk Johnson, Kid Ory, and countless others had been brought out of retirement, Chicago-style jazz had been revived, and traditionalists all over the world could cling to "jazz" without having to entangle themselves with the young rebels in the "bop" and "modern jazz" movements.

The "bop" revolution of the early 1940s had been a long time in fermenting. Many strands led to the eventual storming of the musical barricades. The new institution of the combo contributed its share. Not only the Goodman Quartets and Sextets and the small groups emanating out of the Ellington orchestra, but Shaw's Gramercy Five, Dorsey's Clambake Seven, and independent groups like the Tatum and King Cole Trios, Clarence Profit's Quartet, and the John Kirby Sextet—all contributed to a revivification of the improvising soloist. Kansas City and the Southwest contributed not only Parker but the guitarist Charlie Christian, tenor saxophonist Lester Young, and the bassist Oscar Pettiford. For Parker and Gillespie, the band led by Earl Hines, one of the major "training centers" for young black musicians in the 1930s, became a haven for experimentation in 1943. And the institution of the jam session pulled it all together in Harlem at Minton's Playhouse, where the new leaders, along with pianists Thelonious Monk and Bud Powell and drummer Kenny Clarke, met in almost clandestine privacy, unbeknownst to the merchants of swing.

Once again, the innovative inspiration came from the blacks, foremost among these, as a catalyst, the genius of Charlie Parker. Indeed, Bop was a protest movement against the white bureaucracy and its exploitation of Afro-American music. It was, of course, also a musical revolution: harmonically, melodically, rhythmically, as well as in the realms of structure and form. Advanced harmonies with implications of bitonality (and even atonality) were superimposed on standard tunes. Rhythmically, the original African predilection for cross-accentuation was further released from its 4/4 metric confinement, the beat often became implicit rather than explicit, and above all, the concept of time was lifted to a new plateau—and a new speed level—where the eighth- and sixteeth-note, rather than the quarter-note, became the basic rhythmic unit. The traditional straight-jacket of eight- or four-bar phrases was also broken asunder, and a new freedom in phrase structure ensued.

Although a few orchestras developed directly out of the Bop movement, notably Gillespie's, Billy Eckstine's, Woody Herman's 1940s' Herds and the far-too-little-known bands of Elliot Lawrence and Gerald Wilson, the end of World War II saw the general demise of the big band as an economically and musically viable institution.

Experimentation continued on other fronts. A group of musicians clustered around the blind pianist Lennie Tristano began fascinating explorations of atonality and classical techniques ("I Can't Get Started"). An awareness, often superficial, of contemporary techniques as experienced in the works of Stravinsky, Schoenberg, Bartók, and Milhaud precipitated new stylistic directions in orchestras such as Stan Kenton's (dubbed "progressive"), Boyd Raeburn's, and later, Dave Brubeck's Octets and Quartets.

In the midst of all this, Thelonious Monk, aided and abetted by the brilliant young vibraphonist Milt Jackson, was quietly experimenting with new, oblique harmonies, pitch clusters, and concepts of suspended time. It was not, however, until ten years later (the mid-50s) that the world—even the jazz world—was to discover Monk as one of its greatest composers.

One approach that had a more immediate follow-up was initiated by arranger Gil Evans and two young composers, Gerry Mulligan and John Lewis. Evans, an arranger for Goodman-alumnus Claude Thornhill, began in the late 1940s to evolve a style that embodied the new harmonic advances of bop and the rhythmic language of Parker, and incorporated the use of French horns and tuba, thus giving the band's timbral palette a low-register, dark-hued coloring that (except for certain initiatives of Ellington in a similar direction) was unprecedented in jazz. The legendary Nonet (in recordings since titled the *Birth of the Cool*) led by Miles Davis in 1949-50 comprised the nucleus of that Thornhill instrumentation. The cool, balanced, almost subdued chamber-music quality of these performances was in part a reaction to the freneticisms of Bop, but at the same time represented a desire to at last bring into jazz sonorities that were well established in twentieth-century classical music but unheard (except again for Ellington) in jazz. Evans's "Boplicity" and "Moondreams," Mulligan's "Jeru," and Lewis's "Move" are among the outstanding achievements of this shortlived group.

The "cool" and "modern jazz" movements were perhaps most precisely defined by John Lewis and the Modern Jazz Quartet—which, apart from its profound musical accomplishments in the last twenty years, can boast the most permanent personnel of any group in jazz history. John Lewis as composer and mentor of the group, with Milt Jackson as its most consistently inspiring improviser, developed a chamber style characterized by contrapuntal techniques (derived from a study of Bach), by a classical refinement in tone, ensemble, precision, fastidious attention to detail, all combined with an expressiveness Lewis had heard in the music of Parker, Gillespie, and Lester Young, with all of whom he had worked as a pianist and superb accompanist. Lewis, like his contemporaries, Monk, Charles Mingus, and George Russell, is foremost a composer and as such is in the distinguished lineage of great jazz composers which began with Morton and Ellington. The Modern Jazz Quartet also proved that there are many ways to swing, be it in the subtle, understated variety espoused by the MJQ, or the more muscular, tensile brand of swing of a Mingus, or the relaxed, springy beat of the Basie band.

But the swing of the pendulum in the early- and mid-50s was definitely in the direction of cool and introverted jazz. And before this trend would run its course, the protagonists of the "cool" (such as the post-Lester Young saxophonists Stan Getz and Lee Konitz, the various West Coast groups led by trumpeter Shorty Rogers and drummer Shelly Manne) had reached what many considered the final step, exemplified by Mulligan's piano-less quartet and by clarinetist Jimmy Giuffre's trio *sans* rhythm section.

But the endless cycle of action and reaction had already set in again. A harder, "funkier" jazz was coming back. Trumpeter Clifford Brown, pianist Horace Silver, and Art Blakey's Jazz Messengers, eschewing the refinements and affectations of "classical" techniques, returned to jazz roots for inspiration, mainly the blues and the music of the black church. Into this setting stepped four saxophonists who were in their own ways to change the face of jazz: Sonny Rollins, John Coltrane, Ornette Coleman, and Eric Dolphy. All four typified a revolt not only against the effeteness of "cool" jazz, but even against the "modern jazz" conventions, a heritage left in some state of confusion by the untimely death of Charlie Parker.

Both Rollins with his prodigious facility and almost encyclopedic knowledge of musical traditions (comparable to a computer memory bank) and Coltrane with his agonizing need to struggle through his musical problems in public—expressing the anguish of the artist as a serious quiet man in a violent, crude, and unquiet world—flirted with the avant-garde. But it was Ornette Coleman who in 1959 seemed to break all links with the past. Free forms, free rhythms, free tonality—in short, "free jazz" was the new revolutionary slogan.

Both Coleman and Dolphy immediately became the controversial eye of a storm of revolution. While American and European intellectuals embraced Coleman, Miles Davis—at the advanced age of thirty-four—was putting Coleman down as a fraud; and others accused

Dolphy of not "knowing his changes." (His "Stormy Weather" with Mingus is the best answer to that charge.)

But the bitter reality was unmistakable: what little audience jazz had left by the late 1950s dropped out almost completely after the arrival of Coleman, when the novelty had worn off, and the avant-garde took over in force. In truth, it was a difficult time for jazz. For every one Coleman, there were ten lesser or no-talents who sought refuge in the anarchy and permissiveness of the avant-garde. When in the mid-60s the rock revolution spread over the land like locusts, jazz as it was once known seemed to be emitting its last gasps. Even the avant-garde could barely retain its small audience, part of it being siphoned off by the Beatles and other rock avant-gardists such as Frank Zappa's Mothers of Invention. Electrified instruments and electronic gimmickry pervaded popular music as its decibel-level rose to ear-piercing extremes. Suddenly the choice for millions seemed to be between perennials like Lawrence Welk or Guy Lombardo on the one hand, and the one-chord, Neanderthal, thumping inanities of thousands of rock groups on the other.

But once again the cyclical pendulum came to the rescue. At this writing, as a new generation of youngsters moves into the jazz forefront, the vacuum left by the rock debacle and the vagaries of the avant-garde is being filled by a much-needed reassessment of the immediate past. Audiences who discovered that the "rock revolution" was more sociopolitical than musical are *re*-discovering jazz, not only for itself, but as, lo and behold, an early ancestor of rock (via Rhythm and Blues). And the lessons bequeathed by Parker, Monk, and Bud Powell, which were largely left unlearned by most of the avant-garde, are being restudied. The either-or conflict between free and structured jazz is no longer at issue; the young players simply do both (as with the earlier controversy over "Third Stream"). They are picking things up where the Parkers, Dolphys, Coltranes, Bud Powells, and Scott LaFaros left things some years ago.

2

The Future of Form in Jazz

Schuller's first article on jazz, written in late 1956 for the January 12, 1957, issue of the Saturday Review of Literature. *Though perhaps a trifle "dated" to the reader of thirty years later—given the advances in jazz since 1956—the article does tackle the perennial problem of spontaniety and improvisation in musical performance (jazz or otherwise) and provides an excellent résumé of the new forms and techniques being explored in the late 1950s. It also predicts the soon-to-come arrival of the "next Charlie Parker"; he did not then know his name would be Ornette Coleman.*

IF THERE IS one aspect of present-day modern jazz that differentiates it from the jazz of even five years ago, it is its preoccupation with new musical forms. Jazz today, with its greatly enriched language, seems to feel the need for organization at a more extended level. Few musicians seem to find complete satisfaction in the procedure so prevalent even a few years ago of wedging a group of generally unrelated "blowing" solos and several choruses of "fours" between an opening and closing theme.

At a time, therefore, when one hears and reads terms such as "extended form" and "free form" almost every day, and because there seems to be very little agreement as to what is meant by these expressions, it might be interesting to examine these new tendencies, to see where they may be leading modern jazz, and to investigate what role composition is beginning to play in a music whose greatest contribution has been a renascence of the art of improvisation.

I suppose the question will be raised: why new or extended forms? Why not continue with the same conventions and forms we associate with the main tradition of jazz? Obviously an art form which is to remain a legitimate expression of its times must grow and develop. As jazz becomes more and more a music to be *listened* to, it will automatically reach out for more complex ideas, a wider range of expression. Obviously too, more complex harmonies and techniques require more complex musical forms to support the increased load of this superstructure. The long-playing record, moreover, has emancipated jazz from its previous three-minute limitation, and the *forming* of tonal material on a larger scale has thus automatically become a main concern of the younger generation.

It would be dangerous, however, if the jazz musician were to be satisfied with complacently reaching over into the classical field and there borrowing forms upon which to *graft* his music. The well-known classical forms—such as sonata or fugue, for instance—arose out of and were directly related to specific existing conditions, musical as well as social; and their effectiveness in most cases has been diminished to the extent that these conditions have changed.

For example, the sonata form, originally based upon the dominant-tonic relationship which governed diatonic music, no longer applies with the same priority in an atonal work. This has been amply proven by the discrepancy between musical material and form in certain Schoenberg works, and by the progress made in this respect by Anton Webern and the young generation of composers following in his footsteps. It has become increasingly clear that "form" need not be a confining mold into which the tonal materials are poured, but rather that the forming process can be *directly* related to the musical material employed in a specific instance. In other words, form evolves *out* of the material itself and is not imposed upon it. We must learn to think of form as a verb rather than a noun.

Experience, moreover, has shown us that the borrowing of a baroque form such as the fugue—the most widely used non-jazz form at the moment—very rarely produces the happiest results. Even when successful, it is certainly not the ultimate solution to the problem of evolving new forms in jazz, mainly because jazz is a player's art, and the old classical and baroque forms are definitely related to the art of composing (Bach's ability to improvise complete fugues notwithstanding). Used in jazz these classic forms can, at best, produce only specific and limited results, but cannot open the way to a new musical order. Jazz, it seems to me, is strong and rich enough to find within its own domain forms much more indigenous to its own essential nature.

The idea of extending or enlarging musical form is not a new one in jazz. By the middle thirties Duke Ellington, the masterly precursor of so many innovations in common use today, had already made two attempts to break beyond the confines of the ten-inch disc with his "Creole Rhapsody" of 1931 and the twelve-minute "Reminiscing in Tempo." The latter work, written in 1935, took up four ten-inch sides, but the Columbia label blithely continued to call it a "fox-trot." Its length, its advanced harmonic changes, its unusual asymmetrically-coupled fourteen and ten-bar phrases aroused angry reactions and cries of "arty," "pretentious" and "not jazz." In retrospect we find that it is a poem of quiet melancholy, evoking that special nostalgia which so consistently distinguishes early Ellington from most of his contemporaries; and we see that it was a small, weak step forward to expand form in jazz. Ellington was simply trying to do two things: 1) to break away from the conventional phrase patterns based upon multiples of four measures (in "Creole Rhapsody" he had already experimented with a 16-bar phrase made up of a pattern of 5 plus 5 plus 4 plus 2): 2) to organize his musical material in a slightly more complex form, at the same time integrating solos within that form so that the entire work would produce a unified whole. The least ambitious but perhaps the most inspired of his large-scale works. "Reminiscing in Tempo," opened up a new vista on the jazz horizon.

And yet Ellington—in those days always years ahead of his colleagues—was to wait a decade to see his early experiments emulated by other musicians. Perhaps the intense commercialism of the swing-era with its emphasis on polish (and too often slickness) led jazz temporarily in other directions; or maybe it was simply that jazz had reached a period of consolidation and gestation. Be that as it may, a new style began to crystallize in the early forties under the influence of Parker and Gillespie, a style which already embodied in an embryonic stage the considerable strides jazz has made in the last fifteen years.

It is impossible within the limits of this discussion to examine all the achievements that have led jazz to its present status. The genius of Charlie Parker, the important contributions made by the Miles Davis Capitol recordings, and the success of the Modern Jazz Quartet in popularizing a musical concept that combines classical organization with conventional jazz traditions—all these have already become a matter of history and require no further emphasis here.

More recently, serious contributions to the freeing of form have been made by an ever-increasing number of musicians. Among them (without attempting a complete listing) are: Teddy Charles (although he says most of what he and Hall Overton are doing is not jazz), Buddy Collette, Giuffre, Gryce, LaPorta, John Lewis, Macero,

Mingus, the Phil Nimmons group in Toronto (whose first LP is soon to be released by Norman Granz), George Russell, the Sandole Brothers, Tony Scott, and many others.

A closer look, however, at some outstanding representative examples may help to give a clearer idea of what solutions in the search for new forms have been found.

One of the more interesting uses of form has been developed by Charles Mingus with what he calls "extended form." I think some of the confusion regarding this term arises from the fact that "extended form" can mean simply *that*—extending form in a general way—but also can mean a more specific idea as envisioned by Mingus. For him it means taking one part of a chord pattern, perhaps only one measure, and extending it indefinitely by repetition until the soloist or the "composer" feels that the development of the piece requires moving on to another idea. Actually, this procedure does not represent a new form as such, since it is simply a stretching or magnifying of a standard pattern. Its liberating possibilities, however, are considerable, as exemplified by Mingus's finest efforts in this direction, "Pithecanthropus Erectus" and "Love Chant" (Atlantic 12" LP 1237).

Jimmy Giuffre made a giant step forward with his "Tangents in Jazz," the full implications of which may not be assimilated for years. Aside from his remarkable musical gift, his concern for clarity and logic, his economical means and direct approach indicate that Giuffre is already one of the most influential innovators in present-day jazz. Excellent examples of his concern for formal clarity, with actually extremely simple means, are his written pieces like "Side Pipers," "Sheepherder," and the moving "Down Home" with its Ellingtonish mood of quiet intensity. Atlantic 12" LP 1238).

"Down Home" makes me think of another earlier masterpiece with the same combination of formal perfection and mature musical sensitivity, namely John Lewis's "Django" (Prestige).

What can be done in terms of integrating musical substance with form is also beautifully illustrated by recent recordings of André Hodeir in France. In two albums, not released as yet in this country, Hodeir not only incorporates some of the most recent compositional techniques of European 12-tone writing, but also indicates through them ways in which original forms can be derived from the very core of a musical idea.

Suffice it to cite one especially felicitous example, "On a Blues." Beneath an evenly sustained tenor solo of some length there appears, at first imperceptibly, a riff, which gradually increases dynamically and orchestrationally until it has overpowered and absorbed the improvised solo. The riff, moreover, is not simply repeated in its original form, but undergoes a gradual transformation, at first by means of changing registers, then by inversion, still later by increasingly complex harmonization, and finally through a kind of harmonic and rhythmic condensation of the original riff into a new shape. This building line of intensity, both dynamically and structurally, gives the piece a unique driving force and makes it swing beautifully.

Another remarkable instance of total musical organization (without sacrificing the essential vitality and spirit of jazz) is George Russell's "Lydian M-1" (Atlantic 12" LP 1229), a swinging piece that moves with relentless drive and "quiet fury," to quote Teddy Charles. In a way which is rare in jazz, the entire piece grows tonally and formalistically out of a nucleus of thematic material which in turn is based on a principle which the composer calls the "Lydian concept of tonal organization."

An eighth-note figure of considerable length dominates the opening, and from this, as we shall see, emanates almost all that is to follow. This figure consists at first of a single repeated note (3a) divided into a ¾ pattern (against an underlying 4/4 rhythm set up by the drums), then breaks out into an ascending arpeggio-like pattern (3b) which sounds all the notes of the scale (in parenthesis) that deter-

EXAMPLE 3

mines the tonality of the work. As the thematic line continues, it descends gradually via a series *of* asymmetrically-grouped rhythmic patterns (constantly shifting combinations of 3/8s and 2/8s) to its original starting point, but now grouped in a 3/2 pattern (4a). (The emphasis on ternary rhythms is obvious.) This pattern extended over sixteen measures provides a sort of running commentary to chordal aggregates in the "horns," again derived from the thematic material by

EXAMPLE 4

combining vertically (harmonically) what had previously been stated horizontally (melodically). Four bars of the original repeated-note figure (3a) provide a bridge to a sort of second aspect of the main theme, this time characterized by a blue-note motive (4b) which, however, still relates back to the original underlying modal scale.

During the course of the composition, unity is achieved through reference to this reservoir of material: a 3/2 pattern by the rhythm section contrasts vividly with a trumpet solo in quarter-note triplets; or chordal accompaniments (4c) retain the modality of the opening by being derived through transposition almost entirely from the blue-note motive (4b); a recurring chord progression that frames the three improvised solos is based on the modal scale (3b); and so on.

Above all, the over-all form of the piece is a direct natural product of its own tonal material, giving the whole a feeling of rightness and completeness which marks the work of art.

Now this high degree of integration—which should appeal to anyone admiring order and logic—is considered by many jazz musicians to be too inhibiting. They claim it limits their "freedom of expression" and they consider such music outside the realm of jazz. There is violent disagreement on this point—not without reason. It is a difficult point, usually beclouded by subjectivism and the intrusion of the ego, and it needs to be discussed.

The assumption that restrictions upon intuitive creativity (such as improvisation) are inhibiting is, I think, not tenable, as is demonstrated by all successful art. A great masterpiece, for example, grows out of the interacting stimulus of the constant friction between freedom and constraint, between emotion and intellect.

Charlie Parker seems to have known this. But he also sensed that his work would have been stimulated to even greater heights by the freedom inherent in a context more complex. The chord patterns of his day began to bore him; he said he knew every way they could be played. Many of his solos were so loaded—even overloaded—with musical complexities and razor-sharp subtleties that the implications of a more complex over-all structural level seem incontrovertible. That he did not live to realize the implications of his own style is one of the tragedies of recent music history.

In this connection, there is another point which needs to be aired. It is very much in vogue these days for jazz musicians to "put down" the classical or "legit" way of playing. They scorn the playing of written music (and therefore also composing), and exalt improvisation beyond all reasonable justification.

This is a delicate subject since it touches the very core of a musician's personality and his reasons for being a musician. The subject thus always arouses a defensive and subjective reaction—on either side. I think, however, certain *objective* facts can be stated regarding this controversy which may help to set things right.

Many jazz musicians claim that the classical musician's playing lacks spontaneity, that it has become dulled by repetition and by the very act of *re*producing music rather than by creating it. Only those musicians who have actually played in a first-rate symphony or opera orchestra under an inspired conductor can know to what heights of collective spontaneity an orchestra can rise. After fifteen years of playing in such organizations, I can personally attest to this most positively. Admittedly it *is* rare, since it depends on many factors. But it does occur, and, I think, with more or less the same degree of frequency with which it occurs in jazz.

Listening to several sets on an average night by an average group at Birdland or at an all-night private session will bear this out. How often does a group *really* swing or *really* communicate at an artistic level? (After all, getting "knocked-out" by the beat of an average rhythm section is not yet communication at a very high level. I may respond to it—and I generally do—but that does not by itself make what I'm hearing great music.)

Moreover, is the batting average of a quartet playing "How High the Moon" for the umpteenth time any higher than a symphony orchestra doing its annual performance of the Beethoven "Eroica"? I humbly suggest that the average jazz musician is not in a position to answer that question since he has seldom, if ever, been to a symphony concert and even more rarely has he caught one of those inspired performances. Furthermore, the Parkers, Gillespies, and Lester Youngs exist in the classical world too, only their names are Lipatti, Szigeti, and Gieseking, and *they are indeed just as rare*. Obviously, I do not mean musicians who can improvise like those first three, but soloists who are as highly trained and sensitive in their job of *spontaneously re-creating* a masterpiece as those jazz-greats are in creating.

The illusion of spontaneous re-creation is a factual possibility, as we all know from great acting performances in the theatre. At its highest level, it is an art as rare and as fine as improvisation at *its* highest level—no better, no worse; just different. If re-creating another man's music authentically and illuminatingly were all that easy, then every jazz trumpet player could play the trumpet part from the "Rite of Spring." Obviously what needs to be reiterated is that both ways of playing are highly specialized and require a different combination of skills.

Improvisation is the heart of jazz and I, for one, will always be happy to wait for that 5 percent which constitutes inspired improvising, but is the average jazz musician prepared to look for that 5 percent in classical music?

As for the purists who feel that the pieces under discussion here—and all those works that seem to be gravitating toward classical or composed music—do not qualify as jazz, one can only say that a music as vital and far-reaching as jazz will develop and deepen in an ever-widening circle of alternating penetration and absorption, of giving and taking.

Actually it matters little what this music is called; the important thing is that it is created and that it represents the thoughts and ways of life of its times. Let the academicians worry about what to label it. Seen in this light, the future of this music—jazz or not—is an exciting one. And a fascinating one, because exactly what shape this future will take will not become entirely clear until the next Charlie Parker arrives on the scene.

3

What Makes Jazz Jazz?

A talk given in New York's Carnegie Recital Hall on December 3, 1983, prior to a concert by "jazz" flutist James Newton, part of a series of contemporary jazz concerts presented by Carnegie Hall. The talk was illustrated with brief recorded examples, listed below at the appropriate places.

I SUPPOSE THAT the title of my talk could be viewed by some as being so elementary and bland as to be almost flippant. Flippancy is not, of course, my intention. For underneath that seemingly harmless title and question, there lies hidden a whole series of interesting subsidiary questions and paradoxes and contradictions regarding jazz and its recognition in this country and a true understanding of its nature.

I do not wish to dwell here inordinately on the paradoxes; I would rather focus on some specific musical aspects of jazz that define and separate it from other musical traditions. But I must nonetheless touch briefly on some of the basic confusions which, still to this day haunt jazz, in the very land where this music was born and created, and which in turn make the question "What makes jazz jazz?"—even among relative sophisticates and connoisseurs like yourselves—not entirely irrelevant.

For the confusions and prejudices surrounding jazz are many. The general public's ignorance of it is a national disgrace. It is a sad fact that this music, which now embraces an almost century-old tradition and which is the only truly indigenous, home-grown American musical concept, is unknown and non-existent for the vast majority of Americans. If I were to attach a statistical figure to that statement, I would have to say that this majority constitutes easily more than 90 percent of our population.

Even those who pretend vaguely to know something about jazz, or claim to listen to it, are more apt to be talking about Guy Lombardo and Lawrence Welk, or are likely to identify jazz strictly with Dixieland. Ellington, Charlie Parker, Thelonious Monk, Eric Dolphy—or our artist of this evening, James Newton—are not likely to be known to this vast majority of Americans, or to ever cross their path.

There is a further irony in these sad realities: that this vast majority, while blissfully ignorant of jazz, is totally addicted to one of jazz's primary derivatives, rock and roll and its more recent offshoots.

Embedded in these large confusions and misunderstandings are such further and only slightly smaller confusions regarding the true nature of jazz, as the basic fact that jazz is an inherently creative music, *not* primarily commercially oriented (however much the commercial musics may steal and borrow from it); that it is essentially an improvised music, that improvisation is and has been always the heart and soul of jazz; that it is generally couched in a rhythmic language based on a regular beat, modified by free rhythmic, often syncopated, inflections, all with a specific feeling and linear conception we call "swing"; and finally that jazz is, unlike many other musical traditions, both European and ethnic/non-Western, a music based on the free unfettered expression of the *individual*. This last is perhaps the most radical and most important aspect of jazz, and that which differentiates it so dramatically from most other forms of music-making on the face of the globe.

How this individualism—so typically American and democratic—manifests itself in very specific artistic musical and technical ways will be the main burden of this talk and taped demonstration.

I would like particularly to dwell on one clear way in which jazz distinguishes itself from almost all other musical expressions, even most folk, ethnic, or vernacular traditions, and that is the way jazz musicians play their instruments, with particular regard to the aspect of sonority, timbre, and tone color.

It seems to me that, except for the swing or rhythmic characteristics of jazz performance, the highly individualized sonority aspects of jazz—the sounds of jazz, as musicians simply call it—are its most obviously distinguishing and memorable surface features. In fact, it could be successfully argued that in jazz the individualization of timbre is much more sharply defined than that of rhythm and swing. Although there are several types of swing, and although in the various historical periods of jazz various kinds of rhythmic phrasing, rhythmic feeling, rhythmic conceptions have been explored, it is also true that such conceptions were then adhered to by the great majority of musicians working at any given time, say during the period of early New Orleans jazz, or during Louis Armstrong's time or Charlie Parker's or Ornette Coleman's—all different in specific traits, but all conceptions of high quality, and all held to by those respective individual generations of players.

There is also the point that at either end of the historical spectrum, that is to say, in the earliest days of jazz (before Armstrong's rhythmic revolution) and now in the last two decades (where often the clearly defined rhythmic focus of jazz has become blurred), in both periods, the purely rhythmic distinctions between and among players are much less pronounced—whereas that cannot be said of timbral and sonority aspects. These have always been and still are today highly personal and specific.

There is one other general point to be made which supports the notion of citing sonoric individuality as perhaps the primary distinguishing feature of jazz playing, and that is that the *similarities* between jazz and "classical" music, let us say, are much greater than the *dis*similarities. The pitch language, the harmonic language, is essentially the same—F♯s and B♭s and 7th chords and unnameable atonal chords. So are the rhythms of jazz and classical music. Although harder to notate in jazz because of their greater subtlety and individual treatment, they are essentially the same in concept, that is, they derive from the same basic concept of dividing certain metric units (4/4 bars or 3/4 bars or whatever) into smaller rhythmic units: 8ths, 16ths, half notes, triplets, etc. Similarly the instruments (in both jazz and classical music) are essentially the same, although the jazz instrumentarium seems in general to be more limited and selective: string instruments, oboes, French horns, and the like are still more or less excluded from the jazz-instrument fraternity, and were in the more distant past not considered proper jazz instruments at all and irrelevant to jazz. Moreover, whatever distinctions between jazz and classical music were discernible in the past, they have become much more blurred in recent years with the simultaneous advances, in both jazz and classical music of avant-garde conceptions and techniques and new instruments, to the point where very often now the old sure-fire differentiations between jazz and classical music can no longer be reliably made. Indeed the music of our artist of the evening, James Newton, represents a fascinating example of today's uni-music (as in unisex) that characterizes most of the territory of contemporary music of today. That is why such labels as "jazz" and "classical," never very helpful or accurate to begin with, are now fairly meaningless and perhaps obsolete.

(*play* "Still Waters"—Anthony Davis, 1982)

Without prior knowledge of what that was and who was playing, it would be very hard to identify that unequivocally as jazz.

Rhythm and swing, then, are no longer the automatically distinguishing features of jazz as they were, for example, in the swing era or the early days of modern jazz. Through all these periods, however, the individual sound of jazz—the individual *personal* sonoric conception of players—and the timbral articulations of jazz have remained at the very core of jazz expression, and are today, when the techniques and concepts have in other respects merged so symbiotically, the only distinguishing feature left, by which to clearly identify a performance as being jazz or jazz-oriented—if that still matters at all.

An interesting historical footnote fits here very well. In the 1920s when classical composers like Aaron Copland in America, Darius Milhaud and Ernst Krenek in Europe, discovered jazz, they were first and foremost captivated by the sounds and effects of jazz, the new brass mutes, the new instrumental techniques, the new instruments like the saxophones or the percussion, then either not in use at all in classical music or not used in the way that jazz musicians were utilizing these instruments. Indeed, these composers were impressed by the emphasis on wind and percussion instruments, and the solo virtuosity of the individual players in what was, after all, essentially a chamber ensemble approach, quite removed from the massive sounds of symphony orchestras with their large string sections. It is also true, alas, that that is about all those composers heard in jazz—the sonoric surface of the music. They did not, for example, realize that they were in most cases—Milhaud was a notable exception—listening to dance bands, not true improvised jazz. As a result they were blissfully unaware of the fact that jazz was a primarily improvisatory, spontaneous form of musical expression.

The sounds of early jazz *were,* of course, fascinating—particularly those previously unheard sounds and effects of trumpets and trombones and wailing clarinets.

(*play* excerpts from "Livery Stable Blues"—Original Dixieland Jazz Band, 1917; "Dunn's Coronet Blues"—Johnny Dunn, 1924; "East St. Louis Toodle-Oo," "Black and Tan Fantasy"—Duke Ellington, featuring Bubber Miley, 1927; "Black and Tan Fantasy"—Duke Ellington, featuring "Tricky Sam" Nanton, 1927)

Brand-new, also, were the collective essemble sounds of otherwise highly individual voices, merged into a new composite sonority and texture as in Jelly Roll Morton's "Black Bottom Stomp"

(*play* "Black Bottom Stomp"—Jelly Roll Morton, 1926)

By the mid-thirties Cootie Williams and Rex Stewart, both in Duke Ellington's orchestra, had developed wholly unique ways of coloring their sound, open horn as well as muted, giving that band, in the trumpet section *alone,* a range of colors unheard of in any symphony orchestra.

(*play* "Concerto for Cootie"—Duke Ellington, featuring Cootie Williams, 1940; and "Subtle Lament" (1938), "Dusk" (1940)—Duke Ellington, featuring Rex Stewart)

Similarly, in the trombone section of Ellington's orchestra there was first and foremost Lawrence Brown—here, first an example of an "impossible-to-play" *glissando.*

(*play* "Ducky Wucky" (1932); "Blue Light"; "Main Stem" (1942)—Duke Ellington, featuring Lawrence Brown)

And then there was Juan Tizol, from an Italian/Puerto Rican band tradition.

(*play* "Caravan"—Juan Tizol, 1937)

And there was "Tricky Sam" Nanton, he of the plunger mute.

(*play* "Old Man Blues"—Duke Ellington—featuring "Tricky Sam" Nanton, 1930)

And yet when these three trombonists all played together, they became as one. Different and individual as their respective solo sounds were, their chameleon-like ability to blend into a single ensemble sonority was truly astonishing.

(*play* "Dusk"—Duke Ellington, featuring the trombone trio, 1938)

There at the end, we had some sounds and timbral mixtures (in trombones and saxophones) that had never been heard before in jazz *or* classical music.

Now another example of incredible ensemble virtuosity, "Braggin' in Brass."

(*play* "Braggin' in Brass"—Duke Ellington, featuring the trombone trio, 1938)

In that example the three trombonists were playing in *hocket* style, one note per player in quick rapid-fire succession.

Around the same time another trombone player developed a unique sound and style, which virtually revolutionized not only trombone playing but lyric melodic playing in popular music altogether for most of the swing era.

(*play* "I'm Getting Sentimental," "Tea for Two" featuring Tommy Dorsey, 1939)

And here is yet another great trombone player, Bill Harris

(*play* "Everywhere," "Fan It"—Woody Herman, featuring Bill Harris, 1945).

I have played for you in the last few minutes examples of just five jazz trombonists—I could play twenty-five more, all different—and none of them sound even remotely alike. Such individualism is not only not sought after in classical music, but actually repressed and frowned upon. There the performing practices center on rendering, in a more or less predetermined sound and style, an already existing composition, rather than extemporaneously creating one in a highly personalized way.

Classical musicians, especially wind players, will at most reflect in a subtle way certain national characteristics. German oboists, for example, sound different from the French, both of whom sound different from American oboists. The same is true of bassoonists, clarinetists, French horn players, timpanists, and any other orchestral non-string instrument you can think of. But these differences are mild compared with those we find in jazz. Ironically, these minor regional and national differences are strictly adhered to in each geographic area, so that all German oboists, for example, featuring a very wide and slow vibrato, apply their sound to *all* music regardless of whether it is German music, conceived with that sound in mind, or French music, *not* conceived with that sound in mind, or old music, or new music, or Viennese classical music of Haydn and Mozart, or Romantic music of Bruckner and Wagner.

Only the rarest and finest instrumental artists in symphony orchestras know how to *vary* their sound to render faithfully the different sonoric conceptions of different composers, say Debussy as compared to Beethoven, or Mozart as compared to Tchaikovsky.

In classical music a "beautiful" sound is that which is deemed fashionable at a particular time and place—and these fashions do, of course, change from time to time, every three or four generations perhaps. In jazz, on the other hand, there is no such thing as *a* beautiful sound. It is up to the individual to create *his* sound—if it is within his creative capacities to do so—one that will best serve his musical concepts and style. In any case, in jazz the sound, timbre, and sonority are much more at the service of individual self-expression, interlocked intimately with articulation, phrasing, tonguing, slurring, and other such stylistic modifiers and definers.

One can look at this sonoric individualism also in a historical perspective. Here in quick succession are seven saxophonists through the years, from the thirties to the present, in a dazzling kaleidoscopic variety of sounds and textures—and uses of vibrato. The use of vibrato—of what kind, how wide, how slow, how intense—or the lack of vibrato—is, of course, also a fundamental way in which jazz musicians *color* their sound. (In fact one could give a whole talk just on the varied individual use of vibrato.) The seven saxophonists are:

(*play* "Body and Soul"—Coleman Hawkins, 1939; "Cotton Tail"—Duke Ellington, featuring Ben Webster, 1940; "Dickie's Dream"—Count Basie's Kansas City Seven, featuring Lester Young, 1939; "Hallelujah"—featuring Charlie Parker, 1944; "Blue 7"—featuring Sonny Rollins, 1957; "Alabama"—featuring John Coltrane, 1963; "Lonely Woman"—featuring Ornette Coleman, 1960)

I could have played for you an analogous historical lineage of the trumpet: Louis Armstrong, Bunny Berigan, Roy Eldridge, Dizzy Gillespie, Miles Davis, Lester Bowie.

Similar comparisons and relationships can be made between any *number* of players on any of the standard instruments of the jazz orchestra, and, of course, between *singers:* from Bessie Smith through Billie Holiday and Anita O'Day to Sarah Vaughan and Betty Carter.

Now, to be sure, sticking mutes into brass bells is not the only way of personalizing and modifying a sound. As we have seen with the example of the seven saxophonists (where no muting was involved), vastly different sounds can be produced with different vibratos, different reeds, but also with *how* the air is blown through the instrument and how much resistance from the instrument the player is looking for: the greater the resistance the greater the possibility of a full, rich, powerful tone. Heavier reeds, bigger instruments, wider bores, all these physical/technical aspects affect the sound a player gets, and it is *his* choice, governed by *his* ear, that determines his selection of the appropriate instrumental equipment. He will also take into account that, the heavier and bigger the equipment, the less agile he will be technically, the more difficult it will be to be virtuosic or to play high notes.

And speaking of high notes, the range that jazz instrumentalists command is another aspect of the music's individualistic tendencies and their fundamental desire constantly to expand and extend the expressive potential of the instrument. In almost every case, it was jazz players who expanded the range of our instruments, in ways, incidentally, that have astonished classical players. Louis Armstrong, for example, took the trumpet from its orthodox high note of (sounding) B♭—standard in all the classical etude books, study material, and literature—first to C, then to D♭ and E♭, and finally in the mid-thirties to high F. No classical player would venture that high in those years. Only the cornet soloists like Herbert Clarke or B. A. Rolfe knew that there *were* such high notes on their instrument, but they would use them only at climactic moments and/or endings of pieces. (High C♯'s, as in Strauss's *Salome* or the high E♭ in Schoenberg's Orchestral Variations, were rare exceptions and were either suffered through by the players or ignored.)

But jazz trumpeters by the late thirties could negotiate that highest range with relative ease, playing entire passages up there. One of the most spectacular high-note displays I know of came from the great late lead trumpeter Al Killian, for many years with Count Basie, Duke Ellington, and Charlie Barnet.

(*play* "East Side, West Side"—Charlie Barnet, featuring Al Killian, 1946)

Let's hear it again. It's not only unbelievable, but all too brief; it goes up to the highest F on the piano, a 4th *above* altissimo high C.

(*play* second time)

The trumpeters got so good at extending the range upward that they challenged the clarinetists to expand their range. The clarinet had been content for years to consider high F its upper limit, and few players ever ventured beyond that. But when Armstrong and a host of his disciples began to take high Fs in easy stride on the trumpet, the clarinetists were virtually embarrassed into pushing on upward ever higher. Artie Shaw became the absolute master of this altissimo range. There are hundreds of examples to choose from, almost always in the service of expanding the expressive range of the clarinet, not merely pyrotechnical wizardry. One of the most beautiful examples from Artie Shaw's late thirties' career is this fine solo on "Star Dust." The point about Shaw's high notes is that he could make them sing like nobody else.

(*play* "Star Dust"—featuring Artie Shaw, 1940)

That's a great trombone solo by another fine individual stylist, Jack Jenney.

On the tenor saxophone, too, an extraordinary opening up of the range has taken place over the decades. Time was when Coleman Hawkins, practically the inventor of the jazz saxophone and certainly its first great artist, struggled to reach a high E♭ on his instrument. In "Body and Soul" in 1939 he managed a beautiful F, one step up. Later, Gs and A♭s, and finally B♭s came. Other younger players took up the challenge from there, and by now the tenor saxophone (and its alto partner) have literally gone through the roof.

Although there is always an element of overcoming a technical challenge—a competitive aspect of playing better, bigger, higher than someone else, if you will, a certain ego-satisfaction—all of these advances have generally been to serve the music better, constantly to probe the limits of its *expressive* potential. I say this notwithstanding the fact that some players—there are always a few—abuse these instrumental techniques and use them for purely acrobatic effect, to dazzle an audience with technical stunts, putting technique before content as an end in itself.

But the true jazz artist is generally unimpressed by technique for its own sake. He is generally trying to tell a story, trying to make a personal statement, trying to say something on his instrument that has perhaps not been said before in just such a way. The true jazz artist is creative not just in the on-the-spot invention of musical statements, that is, *his* choice of notes, shapes, phrases, rhythmic figures, but also creative in the way he approaches his instrument. He does this often in ignorance or defiance of the orthodox training methods by which classical players learn their instruments. Many jazz players, especially in the older days, were self-taught. I'm sure it never occurred to Louis Armstrong to stop at high C because that's where some exercise book left off. Similarly, it never occurred to him to use one kind of tonguing or articulation, as most classical trumpet players do. He used whatever tonguing and articulation his ear commanded him to use and in the process invented articulations, phrasings, and rhythmic configurations never heard before.

Armstrong's great cadenza opening of his 1928 "West End Blues" could never have been composed for a trumpet player by a writing-down, sitting composer. It had to be created by a trumpet-playing creative musician who, out of a combination of intuitive feelings, aural imagination, and disregard or ignorance of the orthodox limitations of the trumpet, created a personal statement that revolutionized jazz, all music for that matter, and along the way sort of reinvented the trumpet.

(*play* "West End Blues"—featuring Louis Armstrong, 1928)

It must by now be apparent that the ways in which jazz musicians use instruments and create on them is only as limited as the limitations of their own musical imaginations—which is to say, virtually limitless, infinite, and unpredictable. Who could ever have guessed that this sound and this musical statement could evolve out of the music of Charlie Parker?

(*play* "Congeniality," Ornette Coleman, 1959)

Although there is today a general retrenchment stylistically, and at the same time a careful widespread review of the innovations of jazz of the last two or three decades (and with it a kind of healthy eclecticism), the one thing that hasn't changed is the way in which jazz musicians hear and use their instruments to personalize their art, and to expand it and keep it free. For them it is not the store-bought sound that comes with the instrument and the mouthpiece; it is still a sound bent to their personal needs and expressions as creative artists, refined to speak *their* language, not necessarily someone else's, shaped by their aural imagination, and drawn from the depths of their soul.

Not everyone can achieve these lofty goals, of course. Right now, for example, we have hundreds if not thousands of Coltrane imitators, and very few saxophonists have been able to free themselves of his influence. In the thirties it was Hawkins who dominated the tenor saxophone; in the seventies and early eighties it is Coltrane. But soon someone will come along—perhaps he is already in our midst—who will be strong and imaginative enough to create a new sonoric world on the saxophone through his own individual sound and style.

I should like to end this discussion not with words, but with music. There is no finer summary statement to be made on the subject we have been considering than a performance by the late Eric Dolphy, whose life and career were snuffed out by premature death in 1964.

I purposely left him out of my earlier saxophone lineup so that I could play him now. And as we listen, we might think of the words of that great philosopher-musician Charles Mingus, who once put it this way: "The creative musician must play the music the way *he* feels it. He must stand alone. Music is a language of the emotions; it is a test of life."

Here is a prime example of how an instrument, in this case an alto saxophone, can be transformed into a medium of virtually limitless expressive power and depth, an instrument of beauty and of anguish, an instrument that transcends its own physical, limited self, and can, better than any words of mine, define "what makes jazz jazz." Let us listen to Eric Dolphy and Charles Mingus in their version of "Stormy Weather," as performed and recorded twenty-three years ago.

(*play* "Stormy Weather"—featuring Eric Dolphy and Charles Mingus, 1960)

4

James Reese Europe

Schuller's entry for the Dictionary of American Negro Biography *(W. W. Norton, New York/London, 1982) on one of the most important figures in turn-of-the-century black music.*

EUROPE, JAMES REESE (1881-1919), band leader, composer, violinist, and pianist. He was born in Mobile, Ala., the son of Henry and Lorraine (Saxon) Europe. His family moved to Washington, D.C., where James studied the violin with Enrico Hurlei, assistant director of the U.S. Marine Corps Band. At the age of fourteen James entered a musical contest in which he was defeated by his sister Mary. He also studied theory and instrumentation under Hans Hanke, formerly of the Leipzig Conservatory of Music, and Harry T. Burleigh. In the early 1900s he and another musician, Ford Dabney, moved to New York City where they met other musicians and composers at a hotel run by a colored man, Jimmie Marshall, on West 53rd Street. He found work in various musical shows and in 1906 was the musical director of *The Shoofly Regiment*, a successful production of Bob Cole and J. Rosamond Johnson. Four years later Europe organized the Clef Club, a union comprising many of the best Negro musicians in New York City.

Europe's Clef Club had its own building, which served as a booking office where orchestras of all sizes and types could be hired virtually day and night. The Clef Club venture was such a success that in 1914 Europe took an orchestra of over 125 Negro musicians to Carnegie Hall. Nothing quite like that had occurred before in New York's musical life, and although critical reactions were mixed, Europe's fame spread throughout the city. Moreover, Europe had stormed a bastion of the white musical establishment and made many members of New York's cultural elite aware of Negro music for the first time. A "Negro Symphony," as it was called, playing in Carnegie Hall—that was a concept and setting that could hardly be disparaged as "primitive" or "lowdown."

With the increased popularity of social dancing in the early teens of the century, Europe established another organization, the Tempo Club, this time offering dance orchestras for hire. This led in 1914 to an association with the famous white dance team of Vernon and Irene Castle. Europe composed new dances for them—notably the turkey trot, the foxtrot, and the "Castle Walk"—and supplied orchestras for the Castles' dance salons, *thés dansants,* and tours. In late 1914 Europe was offered a Victor recording contract, one of the very first offered a Negro musician.

When the United States entered the war in 1917, Europe was asked to organize an all-Negro band for the 369th Infantry Regiment, nicknamed the "Hell Fighters." He built the band instrument by instrument, auditioning Negro musicians from all over the country, and conducted it in Spartansburg, S.C., and later on a tour of duty in France. Soon after the arrival of the 369th Regiment at Saint-Nazaire, General Pershing ordered the band detached to entertain troops in leave areas. After playing in several cities and towns, the band arrived at Aix-les-Bains on Feb. 15, 1918. It played not only for troops but also to enthusiastic townspeople there and in nearby Chambéry, the rest station for Negro troops. The band left on March 17 to rejoin its regiment at the front near Givry-en-Argonne. In early September 1918 the band was ordered to Paris for a week of concert duty. It played on Christmas day near Belfort and at Brest on the eve of the regiment's departure from France (Jan. 31, 1919).

The 369th Infantry Regiment, with Europe and his band of sixty pieces of brass and reeds, and a field-music section of thirty trumpets and drums at the head, was the first unit of troops to march through the nearly completed Victory Arch in New York City, erected by the city at Madison Square Park on Fifth Avenue and 25th Street. To the surprise of many, "there was no prancing, no showing of teeth, no swank; they marched with a steady stride, and from their battered tin hats eyes that had looked straight at death were kept to the front" (James Weldon Johnson, *Black Manhattan* (1930), p. 236). The regiment continued north to Harlem where it disbanded.

After the war Europe formed an orchestra and embarked on a triumphant American tour, starting in New York's Metropolitan Opera House. But Europe's dazzling career came to an abrupt end when at the age of thirty-eight, during a concert in Boston's Mechanic Hall, he was stabbed to death by Herbert Wright, a drummer in his band. Lieutenant James Reese Europe was buried with full military honors in Arlington National Cemetery.

Europe's significance to the musical life of America, and particularly to achievements in Afro-American music, has been unduly neglected by historians. Jazz cognoscenti have overlooked him under the mistaken impression that Europe had no relationship to jazz. White and/or "serious music" historians have disregarded Europe because he was considered no more than a popular entertainer. Thus he has almost always fallen between the proverbial two stools.

Through his Clef and Tempo Club leadership, however, he was the first to bring prestige and some degree of professional order to Negro musicians' lives in New York. Moreover, he established his "symphony" orchestras without compromising the essential character of Negro music. He was remarkably lucid and unequivocal on this question: "We colored people have our own music, that is the product of our souls. It's been created by the sufferings and miseries of our race. Some of the melodies we play were made up by slaves, and others were handed down from the days before we left Africa. We have developed a kind of symphony music that lends itself to the playing of the peculiar compositions of our race" (a newspaper interview quoted in Samuel B. Charters and Leonard Kunstadt, *Jazz: A History of the New York Scene,* 1962).

When Europe turned from this "symphony music" to a frankly more popular dance music (in association with the Castles), he was making a decision in response to a dilemma which Negro musicians had always faced (and still do to some extent today): trained as a "serious" musician, should he pursue a career in "classical" music, in the process compromising or abdicating his Afro-American heritage and risking the disfavor of the public by playing music that Negroes were not expected to touch; or should he turn to a career as a "popular entertainer," in accordance with the role the public expected Negro musicians to perform. Europe chose the latter, bringing to his endeavor not only the skills of his earlier training, but a high sense of integrity regarding the role that Afro-American musical elements played in his music. By these means Europe developed a musical style which exhibited an extraordinary rhythmic excitement and exuberance. It is no wonder that Europe enjoyed a phenomenal success. Indeed, his "syncopated music" almost singlehandedly influenced the musical tastes, the social and leisure life of virtually an entire generation of Americans. The electrifying momentum and rough excitement of Europe's style can fortunately still be heard on his recordings (for Victor and Pathé) which, although hard to find, are worth the search.

Europe's vision of music drew from both his "classical" training and his own ancestral heritage. In both his Society Orchestra and the "Hell Fighters" Band, he demanded (and got) from his musicians the precision and control associated with "strictly written" music, combined with a new improvisatory freedom and directness. Already present in the best of ragtime, Europe's translation of these elements to the orchestra and the band multiplied the potential musical impact. Symptomatic of the sense of abandon such music could generate are the "breaks" taken by the entire clarinet section of the "Hell Fighters" Band in Europe's recordings of "That's Got 'Em" and "Clarinet Marmelade."

Europe was, along with "Jelly Roll" Morton, the most important figure in the prehistory of jazz. And like Morton he added new rhythmic dimensions to ragtime and prepared the way, especially in New York and on the East Coast, for the full emergence of jazz. By the end of his life he was enjoying a popular success which can only be described as phenomenal and which was unequalled by any Negro up to that time.

Had he lived, he would have undoubtedly achieved even greater worldwide success and might have gone on to make a counterpart to Morton's Red Hot Peppers or Armstrong's Hot Five recordings. Failing that, he nevertheless was a figure of major significance, for he brought a new dignity to Negro musicians. And as the leader of a kind of Negro musical avant-garde, he gave a new thrust to the development of Afro-American music. He was the real initiator of the Jazz Age and a mentor of countless numbers of Jazz Age musicians.

Europe died in a Boston hospital on May 9, 1919. Funeral services, attended by John Wanamaker, Sr., Maj. Hamilton Fish, Jr., members of the "Hell Fighters" Band and of the 369th Infantry, of the Hayward Unit of the National League for Women's Service, Masons and Elks, were held at St. Mark's M. E. Church, New York City, on May 13. Harry T. Burleigh sang "Victory." Interment with full military honors was in Arlington National Cemetery. He was survived by his widow, Mrs. Willie Europe, a mother, sister, brother, and other relatives whose names were not given in the *New York Age* (May 17, 1919, pp. 1 and 6). It is known, however, that his sister, Mary Europe, taught music in the public schools of Washington, D.C.

For Europe's career and significance in the early 1900s, see James Weldon Johnson's *Black Manhattan* (1930, esp. pp. 119-24). Photographs of the Clef Club Band, Europe conducting (1914); the Clef Club Orchestra (1915); a program of the Tempo Club presenting Mr. and Mrs. Irene Castle (1915); of the "Hell Fighters" Band marching in France (1918), returning on board ship (1919), and the parade on Fifth Avenue are in *Harlem on My Mind, Cultural Capital of Black America,* edited by Allon Schoener (1968, pp. 42-43, 46-47). There are other photographs in the best source for his experiences in Europe, Arthur W. Little's *From Harlem to the Rhine, The Story of New York's Colored Volunteers* (1936, betw. pp. 158 and 159, and on pp. 174, 214, 215). Irene Castle's "Jim Europe—a Reminiscence" (*Opportunity,* March 1930, pp. 90-91) is a warm tribute.

5

The Orchestrator's Challenge

A brief introduction to Schuller's approach in orchestrating Scott Joplin's opera Treemonisha. *The opera, which was produced by the Houston Grand Opera Company, was later taken to the Kennedy Center and Broadway. Subsequently it was issued by Deutsche Grammphon in an original-cast recording (2-DG 2707083), and it was for the latter's program book that Schuller's comments were written.*

THE PROBLEMS FACED by the would-be orchestrator of Scott Joplin's *Treemonisha* are very special ones, for *Treemonisha* is a very special opera—one of those wondrous *sui generis* creations which our musical culture occasionally produces. The result of a three-way cross-breeding of elements—mid-nineteenth-century European opera, Afro-American dance forms, and turn-of-the-century American popular (or, as they used to call it, semi-classical) idioms—*Treemonisha* is a curious alchemical mixture of musical styles and conceptions which would certainly have failed in the hands of a lesser talent than Joplin's.

Thus, the "philosophy" underlying an orchestration of *Treemonisha* must pay equal respect to each in this remarkable tripartite confluence of styles. Furthermore, it must accept and be sensitive to the twin facts that the music of *Treemonisha* is a wonderful *anachronism,* some exceptionally "advanced" moments notwithstanding, and is—the most difficult of all elements to recapture—a period piece full of innocent charm of a kind that our late-twentieth-century world has long ago lost but is evidently intent on regaining. And it goes without saying that one must clearly understand the fact that *Treemonisha* is not a "ragtime" opera. Finally, one must know the kinds of things Joplin might have done with instruments and, perhaps more importantly, what he would *not* have done. All of this, it must be clear by now, precludes any musical "updating" of the work. I see it strictly as a challenge of *re-creation.*

With all these impositions, the path I trod was a narrow one, yet like some craggy mountain path, splendidly challenging. I chose an orchestra of about thirty-five instruments, which represents a balance between what Joplin, the visionary, might have dreamed of for his opera and what Joplin, the realist, would have been forced to accept, i.e., between an economy-sized romantic opera orchestra and a pit-orchestra typical of the kinds of vaudeville, variety show, and theatre–Broadway musical formations which were in vogue during the first two decades of the century. The orchestrations of the "Red Back Book" from around 1910 furnish some excellent clues: a mini-symphony orchestra containing the four woodwinds (flute, clarinet, oboe, bassoon), the four brass (trumpet, horn, trombone, tuba), a small string section (whatever one could afford or whatever the pit would hold) and a rhythm section (in those days rarely with banjo or guitar). The saxophone was also a later addition, belonging more to the jazz age, just *after* the writing of *Treemonisha* and Joplin's death in 1917. By 1915, such orchestras would have grown to include *pairs* of clarinets, trumpets and horns. The orchestra for Eubie Blake's 1921 "Shuffle Along"—the first Negro musical to be a smash hit on Broadway, setting the tone for future jazz-oriented shows—had such an enlarged pit complement. (The dean of black composers, William Grant Still, was the oboist of that "pit band.")

Lastly, what *kinds* of parts Joplin would have written for those instruments must be viewed in the context of the anachronistic nature of the work. When Joplin finished *Treemonisha,* Strauss had already stunned the world with his *Salomé* and *Elektra,* Schoenberg had just composed *Erwartung* and Bartok his *Bluebeard's Castle.* But Joplin was totally unaware of such orchestral avant-gardism. Thus Straussian horn parts, virtuosic string writing, or sophisticated percussion parts are totally foreign to Joplin's world. To re-create Joplin's music authentically, one must resist the temptation to let the "modern" twentieth-century orchestra of Ravel and Stravinsky, and even latter-day Broadway, creep into the work; one must hew to the venerable, tried and true idiomatic formulas which dominated Joplin's music-theatre milieu. It is this very "quaintness" of instrumental writing which is the ideal counterpart to the innocent and simple faith that forms the essence of Joplin's operatic fable.

6

Happy Feet: A Tribute to Paul Whiteman

Many who had typecast Schuller as an "avant-gardist" and radical "serialist" and "abstractionist" were surprised—and perhaps even dismayed (because it spoiled the perceived image)—when he espoused such artistic causes as the music of Scott Joplin and Paul Whiteman. This article appeared initially as the liner-note commentary for Schuller's re-creation with students of the New England Conservatory of fifteen Whiteman recordings from the late 1920s (on Golden Crest 31043). It was, in turn, an expansion of a by-now famous footnote on Whiteman in Schuller's Early Jazz.

IT IS UNLIKELY that anyone in jazz history has been as consistently maligned as Paul Whiteman. This recording hopes to put Whiteman's contribution to jazz in proper perspective and, by paying tribute to him, to right a wrong which has been perpetuated far too long.

For decades jazz writers have propagated the myths that, except for Bix's (and a few of Tram's) solos, none of the Whiteman recordings—of which there are hundreds—are worth hearing; that Bix suffered intolerably in the Whiteman band; and that Whiteman's concept of "symphonic jazz" and his attempts to "make a lady out of jazz" were a betrayal to the music. A favorite target of jazz writers has been the use of violins, for this is where the real betrayal of jazz is alleged to have occurred.

Bosh! to all of that. The Whiteman orchestral sound, as fashioned by three of the greatest arrangers that ever worked in the field—Ferde Grofé, Bill Challis, and the Dean of Black American composers, William Grant Still—is one of the half-a-dozen most original jazz sounds ever to have been created, and it provided the perfect setting and inspiration for Bix: like a jewel set perfectly in the center of a crown.

In its own way the Whiteman sound is as original and as beautiful as that of Duke Ellington's orchestra—very different, of course, but no less magical, no less inspired. And let no one forget that Ellington and the other great jazz orchestra leaders of the late twenties spent a great deal of time learning from the Whiteman "book," emulating it and marveling at its instrumental sophistication.

Part of the beauty of this sound derives from the fact that it makes a much greater than usual use of the full range of "orchestral" instruments beyond the traditional jazz band of the twenties: a nine-piece brass section, a six-man reed section that played as many as twenty-five different instruments (from E-flat clarinet to Heckelphone), variously sized violin sections, a wealth of percussion instruments, banjo, and (often) two pianos and celesta. What is especially remarkable about this instrumentation is that it allows an arranger, if he so chooses, to cover virtually the entire six-octave range of our modern orchestra as it developed in the late nineteenth century—from the lowest notes of the tuba and bass up through the orchestra with (in Whiteman's case) no gaps right up to the highest register of the violins and piccolos.

And how Challis and Grofé and the others used those possibilities! They were far ahead of their time in all respects: harmony, rhythm, timbre, technical sophistication, and in their best work, perfection of form. Moreover, they had an incredible group of musicians to deliver the goods—from the incomparable Chester Hazlett (reeds) through Matty Malneck (violin and viola) and Mike Trafficante (tuba) to Charlie Margulis (trumpet), not to mention Bill Rank, Tommy and Jimmy Dorsey, Joe Venuti, Eddie Lang, Frankie Trumbauer, and, of course, Bing Crosby and Bix Beiderbecke. Whiteman demanded musical discipline and perfection, and nearly always got it. It is not for nothing that Darius Milhaud in his autobiography wrote of the unprecedented perfection of the Whiteman orchestra when he first heard it in Europe in the twenties, likening it to a "well-oiled Rolls Royce."

This, then, is a tribute to Paul Whiteman and the remarkable music that he caused in one way or another to be created. It is also a record which should provide much aural pleasure and remind us what great songs were written and played in the twenties. In some ways, the excellence of the Whiteman *repertory*—admittedly selective—will perhaps be the biggest surprise of all to many listeners: one great unforgettable tune after another. And for those who cherish the "verses" of songs as well as the "choruses," there will be many delightful moments on this record, as Grofé and Challis time and time again ingeniously integrate the verse into their arrangements.

The authenticity of these performance re-creations of nearly fifty years ago was immeasurably aided by the fact that we were able to use the original scores and parts, kindly made available to us by Williams College, the proud owner of the entire Whiteman Collection. Our efforts were centered on re-creating these remarkable performances as nearly accurately as possible, including Bix's solos—played here by Bo Winiker—but not including the numerous inimitable vocals by Bing Crosby and the Rhythm Boys. A number of the humming and crooning ensembles lent themselves rather well to transcription for cup-muted trombones; similarly Bing's infectious scat solo on "Changes" is played on the trumpet by Winiker.

To anyone who will listen to these performances with an open mind (and open ears), it will be abundantly clear that Paul Whiteman and his orchestra of the 1920s comprises one of the treasure chests of our American musical heritage.

7

Ellington in the Pantheon

A tribute to Duke Ellington written shortly after his death for the November 1974 issue of High Fidelity. *As he had done often before, Schuller here, too, suggests a rightful place for Ellington the composer in the larger context of music, regardless of category—not just "jazz," a restrictive connotation Ellington himself rejected.*

WHAT IS THERE left to say about the art of Duke Ellington after a lifetime of successes caressed in superlatives and now, since his death, after months of I-knew-him-too tributes by musicians and fans alike?

Very little, I suppose—except that as usual, and perhaps understandably, much more attention has been given to the man, the charismatic Ellington personality, the inveterate traveler of thousands of one-night stands, Ellington the tune writer, than to his compositions. Admittedly, it is hard to talk about music in words: Music, especially Duke's music, speaks better for itself, and talk *about* music is often necessarily subjective and impressionistic. On the other hand, there are some things to be said about all great music that are more objective and factual than we sometimes care to admit. For greatness is not altogether accidental, altogether intuitive or mysterious. Much of it results from simple hard work, selflessly applied energy, and a fierce determination to learn and apply what has been learned.

If I dare to include Ellington in the pantheon of musical greats—the Beethovens, the Monteverdis, the Schoenbergs, the prime movers, the inspired innovators—it is precisely because Ellington had in common with them not only musical genius and talent, but an unquenchable thirst, an unrequitable passion for translating the raw materials of musical sounds into his own splendid visions. But that is still too general, something that can be said even of minor composers.

What distinguishes Ellington's best creations from those of other composers, jazz and otherwise, are their moments of total uniqueness and originality. There are many such flashes in his *oeuvre,* and it is a pity that they are virtually unknown to most non-jazz composer colleagues. Perhaps this is due to the fact that you cannot go into the nearest music store or library and obtain the orchestral scores of Duke Ellington. There is no Ellington *Gesamtausgabe,* alas, although this is something that should become someone's life work. However, even if such scores existed, they still would not readily disclose the uniqueness of which I speak. For Ellington's imagination was most fertile in the realm of harmony and timbre, usually in combination. And as played by some of the finest musicians jazz has ever known, the specific effect produced in performance and on records is such that no notation has yet been devised to capture it on paper.

Nevertheless they exist—alas *only* on records, and they are none the less real for that and no less significant. The opening measures of "Subtle Lament" (1939) (Ex. 1), and the second chorus of "Blue Light" (1939) (Ex. 2)—both wondrous harmonic transformations of the blues; the muted brass opening of "Mystery Song" (1931); the

last chorus of "Azure" (Ex. 3a, 3b) with its remarkable chromatic alterations; or the total orchestral effect of the first bridge of "Jack the Bear" (1940) (Ex. 4), not to mention the uniquely pungent harmonies of "Clothed Woman" (1947): These are all moments that can literally not be found in anyone else's music. They are as special and original in their way as the incredible D minor-D sharp minor mixture and instrumentation that opens the second part of the *Rite of Spring* or the final measures of Schoenberg's *Erwartung.*

Citing musical examples can give only a severely limited impression of the total effect in performance. For finally it is the unique sound of a "Tricky Sam" Nanton, a Cootie Williams, a low-register Barney Bigard that transmutes those harmonies into an experience that even master colorist/harmonists like Debussy and Ravel could not call upon from their orchestras.

It was part of Ellington's genius—what I called earlier his fierce determination and unquenchable thirst—to assemble and maintain for over forty years his own private orchestra, comprising musicians more remarkable in their *individuality* than those of any symphony orchestra I know. Not since Esterhazy had there been such a private orchestra—and Esterhazy was not a composer. But like Haydn, who practiced daily on that band of Austrian/Hungarian musicians to develop the symphonic forms we now cherish, so Ellington practiced on his "instrument." This is a luxury we other composers simply do not know, and the whole experience of writing consistently for a certain group of musicians is a phenomenon we have never savored.

In Ellington's case, collaboration of such intimacy and durability was bound to produce unique musical results. These can be heard on literally hundreds of Ellington orchestra recordings in varying degrees of "uniqueness." When that alchemy worked at its best, the result was such as cannot be heard anywhere else in the realm of music.

A large statement? Preposterous? Check it out for yourself. The originality of Ellington's harmonic language, with its special voicings and timbres, gives the lie to the often-stated suggestion that he learned all this from Delius and Ravel. Rubbish! This is no more tenable than it is to say that Debussy and Ravel sound alike, even if they both use ninth chords. Like these masters, and others such as Scriabin and Delius, Ellington always found a special way of positioning that chord, of spreading or concentrating it, of giving it a unique sonority that cannot be mistaken for any other's.

Like Webern, he limited himself to small forms—a few notable exceptions notwithstanding. In fact it was not entirely by choice in Ellington's case, but the three-minute ten-inch-disc duration was simply imposed on jazz musicians for a variety of technical/practical/commercial/social/racial reasons. What matters is that he took this restriction and turned it into a virtue. He became the master in our time of the small form, the miniature, the vignette, the cameo portrait. What Chopin's nocturnes and ballades are to mid-nineteenth-century European music, Ellington's "Mood Indigo" and "Cotton Tail" are to mid-twentieth-century Afro-American music.

In his inimitable way the Duke towered over all his contemporaries in the jazz field and equaled much of what is considered sacred on the non-jazz side.

He is gone now, alas. Yet his music lives on and is still with us—at least on recordings. I believe that is not enough.

8

Ellington vis-à-vis the Swing Era

An assessment of Ellington's studio recordings for the year 1938, written for the Smithsonian Institution's multi-record Ellington anthology and covering the years 1938 through 1940, considered by most to be the peak period of Ellington and his orchestra.

WHEN 1938 arrived, Duke Ellington and his favorite "instrument," his orchestra, had just survived a rather troubled two years. Apart from various personal and family worries, there was the Great Depression still to contend with, a dramatic wrenching of the social fabric of our country and in many ways both the end of an era and the beginning of a new one. Since all musical evolution is a reflection of social changes, indeed a creative parallel to them—and in no music is this more true than in jazz—these new social tremors were very much felt in the jazz and popular music of the time. It was the beginning of the Swing Era, and indeed in the memory of many now middle-aged Americans the mid-1930s, the Depression, the New Deal and Swing are all virtually synonymous.

Ellington and his musicians were not impervious to these pressures, although it must be said that they survived rather well compared to a sizable number of other jazz orchestras founded in the 1920s which did not manage to outlive the Depression. Within the orchestra, personnel changes were beginning to affect its performance. After an unusual decade of stability, the situation was beginning to fluctuate as illnesses and various kinds of dissatisfactions began to plague the band. Toby Hardwick and Barney Bigard were absent at various times; Arthur Whetsol, one of Duke's irreplaceable mainstays, was suffering more and more from the cancerous condition which was to take his life in 1940; and Freddy Jenkins, another trumpet stalwart, suffering from tuberculosis, had left the band in 1935. Of course, there were compensations and replacements for these losses. Rex Stewart, a formidable musical personality, had joined Duke in late 1934. To replace bassist Wellman Braud, Duke found himself with two strong players, Billy Taylor and Hayes Alvis. There were rumors of other departures, and some wags even had the band breaking up. Then there was the increasingly carping comments of the critics who were unable to deal with Duke's originality, his frequent excursions into new forms ("Reminiscin' in Tempo," "Diminuendo and Crescendo in Blue"), new compositional experiments ("Daybreak Express," "Azure," the series of "concertos" for Barney Bigard, Cootie Williams, Rex Stewart, and Lawrence Brown), and in general Duke's resistance to typecasting and to being cast into those molds which were then prevailing in jazz.

By 1938 that prevailing jazz mold was *swing,* swing as the popular name for a style. Jazz in the form of swing had in fact become *the* popular, entertainment and dance music of the U.S. Ellington, always very much his own man, was not about to be crowded into a narrow stylistic corner. Besides, there was pungent irony in the fact the Duke and many other fine black musicians had been using the word swing as a way of playing jazz, indeed a mandatory element of jazz, for years. Ellington's "It Don't Mean a Thing, If It Ain't Got that Swing" from 1932(!) had been but one particularly innovative manifestation of that reality.

Now in the mid- and late 1930s, when everybody else had finally caught up with swing, both as a noun *and* a verb, Duke resented being stereotyped by something he had accomplished years before, and had long since transcended.

But swing music was making headlines everywhere, and Ellington was simply considered part of the swing movement, whether he liked it or not. His performances were called "swing concerts," college campuses began to take note of the Ellington band, and as the swing craze invaded Hollywood, Duke and his men were presented as leaders in the field. Suddenly the depressing and uneasy years of the mid-1930s gave way to a new sense of prestige for Duke. For not only was he considered a leader in the vanguard of swing and jazz, but as a veteran bandleader of a dozen years or more, Duke now found himself garnering a level of respect from audiences as well as colleagues that was new in jazz history.

Inevitably, even Ellington was unable to resist altogether the commercial, financial, and psychological pressures—in short, the pressures of the market place—in a society demanding swing music for its pleasure and consumption—and by God, it was going to get it! The year 1938 sees Ellington struggling with these pressures—sometimes unsuccessfully. But in the end—particularly if we look beyond 1938 to the later fruits of this struggle, the superb Ellington masterpieces of the early 1940s—Duke remained true to his art, to his people and to himself.

If the wild success of "swing music" was bound to occasionally assert itself in Ellington's music at this time, it is perhaps inevitable that the first title the band recorded in 1938 was called "Stepping into Swing Society." And in truth, there seems to be—for the moment—a new spirit in the orchestra, a fresh feeling which manifests itself in a variety of ways. There is the added big-toned voice of Harold "Shorty" Baker in the trumpet section; a new springy lift in the ensemble playing; a confidence not heard in the band for some time and exemplified best by Carney's ebulliently imperious solo; and an overall fresh sense of swing and momentum, nourished by Sonny Greer's inspired brush work. It was as if a curtain had been rung down on the agonizing frustrations of the past thirty-six months. The torment and anguish one hears in "Diminuendo and Crescendo in Blue," "Reminiscin' in Tempo" seemed now to give way to a new clarity and economy.

With this passing obeisance to the swing craze out of the way, Ellington returned to that early quintessential masterpiece, "Black and Tan Fantasy" (1927). Although idiotically issued on two separate sides on the Brunswick label as "Prologue to Black and Tan Fantasy" and "The New Black and Tan Fantasy," Ellington's reworking of this classic was an expanded "recomposition," a single work split into two parts by the mandatory limitations of the ten-inch, three-minute-plus shellac disc to which jazz music—even Ellington's—was still relegated in the 1930s. Much of the earlier work remains intact, including Bubber Miley's original inspired plunger-and-growl trumpet solos, becoming after Miley's death the virtually exclusive property of Cootie Williams. As used in the new Cotton Club show for which the Ellington band provided the music in 1938, the piece needed to be expanded and thus there are longer solos for Cootie as well as newly interpolated solo choruses for Barney Bigard and "Trick Sam" Nanton. Among the many new highlights of the performance are the sombre introduction in fifths by the two bass players, followed by the collective "growling" of the entire brass section, Cootie's long swelling high B-flat—a standard, attention-getting device ever since Louis Armstrong's 1928 "West End Blues," but still always thrilling—and finally Bigard's hair-raising twelve-bar-long single note (in one breath!), bent in an agonizingly slow *glissando* from a high D-flat up a third to F. The latter is surely an example of how a tawdry cliché could be used with taste, dignity and beauty, for the clarinet glissando and high-register squeal were thrice familiar tricks of the trade ever since the likes of Ted Lewis, or on another level, the famous opening clarinet *glissando* of Gershwin's "Rhapsody in Blue."

The next three sides—"Riding on a Blue Note," "Lost in Meditation," "The Gal from Joe's"—are vintage Ellington of the period, the first a Cootie Williams vehicle in which he stretches for the first time for a high F (then considered the just barely attainable outer range of the trumpet); the second a poignant ballad played with sovereign suavity by Juan Tizol on valve trombone; the last a Johnny Hodges piece in a minor key, a riff nmber thematically on the weak side, but still easy grist for Hodges's creative mill.

This may not be the place to argue the question of chronological primacy, but suffice it to say that the famous Glenn Miller reed section sound, first exhibited by the Miller band in 1939, is here present, full-blown, on Ellington's "Lost in Meditation," recorded February 2, 1938!

"Skrontch" (also spelled "Scrounch") was the Cotton Club Parade's big dance finale in which Ivie Anderson exhorted everyone to learn this new dance. The song practised what it preached, featuring an unexpectedly strong accent on the fourth beat:

4/4 Skrontch on the fourth beat; Skrontch etc.

Also written for the Cotton Club show, and of a higher order, was the fine torch song "I Let a Song Go Out of My Heart." It was not only another superb vehicle for the artistry of Johnny Hodges, but in its last sumptuously velvety twelve measures, it predated and presaged a number of other fine moments in subsequent jazz history: Duke Ellington's own hit of 1940, "Don't Get Around Much Anymore"; Glenn Miller's "In the Mood" ending; some of Count Basie's 1939-1940 hits; and much later, the sound that Gil Evans made into the trademark of the later 1940s' Claude Thornhill orchestra.

The Cotton Club Parade's most stunning instrumental showstopper must have been "Braggin' in Brass," a tour-de-force of unsurpassed brass virtuosity—particularly by the trombone section—still making players *today* shake their heads in disbelief. The basis for the piece is the old standard set of harmonies used in "Tiger Rag," already in that earlier guise a virtuoso display piece. Rex Stewart, the main soloist, resuscitated his earlier concerto vehicle "Trumpet in Spades," a fast melange of ancient "cornet solo" clichés and pre-bop eighth-note runs, topped by grandstanding high E-flats and Fs. This somewhat thin and shopworn material, however, was greatly enhanced by Ellington's addition of a whole new first section, featuring the three trombonists in *hockuet* style at such amazing speed that one's first reaction is, in fact, disbelief. "Hocket style" refers to a basically polyphonic concept of performing in which a total pattern is created out of single interlocking notes or note-groupings. (Trinidad steel drum bands use this technique, and it has been a centuries-old tradition not only in European early Renaissance music, but also such musics as far removed as West African ivory horn ensembles.) In "Braggin' in Brass"—what an apt title!—each of the three trombonists has one note assigned to him in a descending triadic pattern, as follows:

the composite of which (in piano score) looks as follows:

The fact that this is done at a murderous tempo of ♩ = about 280 and that the triadic groupings form a 3/8 pattern, overlaid on the basic 4/4, makes the over-all result all the more stunning. Lawrence Brown's blistering solo later on is another measure of this great artist's near-perfect control of ideas and technique.

The same two artists (Brown and Stewart) team up again as the primary soloists in the next offering, "Dinah's in a Jam," a typical late 1930s swing riff piece with a distinctly small-band flavor.

By mid-1938, the Cotton Club tunes had all been recorded and Ellington turned to other compositional interests. Unlike most bandleaders and composers or arrangers, he delighted in creating in *several* musical categories: not only the expected ballads and dance numbers that are the bread and butter of any jazz band, but also "production numbers" for shows (like the Cotton Club shows), the blue or "mood" piece (like "Mood Indigo") that as a category was really invented by Ellington; pieces which were simply abstract "musical compositions" (as those of classical composers, but instead of calling them Etudes or Ballades or Partitas or Symphonies, Ellington would give them ingenious descriptive titles); and finally, the category of pop tunes.

"You Gave Me the Gate" was one such latter endeavor—not particularly inspired or successful at that. It updates the old minstrel hokum routines of the turn-of-the-century and plays on a swing-conscious generation's growing awareness of the Negro jazz musicians' "hip" lingo. "You gave me the gate, and I'm swingin'," Ivie Anderson opines. Its compromise with well-worn success formulas shows even in Rex Stewart's (perhaps-tongue-in-cheek) rendition of a Harry James-type solo, the kind with which the Goodman band was burning up the countryside about this time.

We have included two takes of "Rose of the Rio Grande," not only because it features impeccable offerings by Ivie Anderson and the great trombonist Lawrence Brown, probably Ellington's most versatile musician, but because it is audible proof that Ellington's soloists often committed "solos" to memory and played them the same way—with but minor deviations—night after night, often over a period of years. Brown's solo is one of his most distinguished statements, a perfectly organized solo, one which became an instant favorite with Ellington audiences. The point is neither that memorizing a jazz solo is bad, as some purist jazz critics have tried to maintain, nor that it is good, as has been argued by those who feel the need to "dignify" jazz by equating it with "classical" music, but rather that, *being* memorized—Brown played virtually the same solo when I first heard him in person six years later in 1944!—it always sounded like fresh improvisation.

"Pyramid" was an attempt to recapture the success of the Ellington-Tizol hit of 1937, "Caravan." But it evolved as rather a bit of a hybrid in which the sum of its parts did not add up to a totality. The "mysterious near-East" is evoked by Tizol on muted valve trombone with Ellington himself playing the hypnotic rhythmic *ostinato* on a homemade hand drum under Greer's snare drum beat. Without preparation Harry Carney bursts into the tranquil scene with a brief jazz statement. A suave trombone trio in shifting D-flat and C harmonies follows, but that mood is shattered by a tawdry full-orchestra section sounding as if left-over from some big-theatre vaudeville act.

"When My Sugar Walks Down the Street" is another discreet bow to the pop market of the day with "birdies going tweet-tweet" and all. But no matter how ordinary the material, the Ellington band with its tasteful rhythm section and battery of virtually infallible soloists always transcends such inherent limitations.

How deeply expressive a vehicle the romantic ballad had become for the Ellington orchestra is exemplified in "A Gypsy Without a Song." A full description, let alone an analysis, is beyond the scope of these notes, but a few of its fascinations must be pointed out. There is, first of all, Ellington's mastery of form as the small form of the AABA chorus structure is mirrored in the large form of the overall performance, creating a complex of interlocking strata of intensity, and, miraculously, something close to a "sonata form." Then there are Hodges's and Williams's sovereign solos, so economical (a lesson to today's thousand-notes a minute players), so self-contained and yet so moving. But perhaps even more fascinating is the first chorus in which Ellington, like a master illusionist, extracts subtleties of color from his orchestral palette that very few (if any other) musical imaginations have envisioned. The melody in the first sixteen bars is split between two trombonists: Tizol on valve trombone, Brown on slide—but so identically muted that on a perfunctory listening one would assume the presence of only one player. This sleight-of-hand trick is managed by interpolating two bars of open-horn trumpet by Cootie Williams in a decidedly non-ballad mood, thus neatly disguising the seam where Tizol's and Brown's eight-bar phrases join. Further, the subtle intensification of melancholia that Brown achieves by his refined use of the slide (over Tizol's "straight" version) not only exploits in a perfectly disarming way the basic idiomatic difference between the two types of trombones, valve and slide, but uses this physical difference as a subtle variant in expression. It also seems to me that Ellington creates the illusion of having extended his instrumental resources by conjuring up a muted French horn (Tizol) and the soulful *portamenti* of a cello (Brown), even though neither of these instruments were present in Duke's orchestra.

It is this kind of musical wizardry and subtlety that has always placed Ellington far above any of his contemporaries.

But as we have seen, such creative peaks were surrounded by lesser mountains, hills and even valleys. The Ellington orchestra in 1938 and 1939 seemed to be in a kind of stylistic holding period, for with some notable exceptions the compositions and arrangements catered to the dicta of the Swing Era. In pieces like "Buffet Flat," "Hip Chic," "Old King Dooji," "Love in Swingtime," "Jazz Potpourri"—as well as numerous Ivie Anderson vocals ("La De Doody Doo," for example)—Ellington appeared to have restrained his creative energies, content to husband his resources and ride out the competitive commercial pressures exerted by the sudden mass proliferation of swing bands and the American public's embrace of swing as its national music. Appropriately decked out with the latest swing stereotypes of unison sax lines and repetition-laden arranged ensembles, sometimes even overplaying the drum parts (as on the 1939 "Weely" and "Way Low"), perhaps in an attempt to deal with a public weaned on Gene Krupa's exhibitionism, these swing riff pieces merely reflected the constraints of the marketplace.

Sometimes the tie-in to the marketplace was quite direct and unmistakable. In 1938 a host of dancers, song pluggers, and music publishers tried to import a Cockney dance step, the Lambeth Walk, into America. As one writer once put it, the dance "died on its feet," even though Ellington's "Lambeth Walk" remains as a testimony of this venture.

But in the midst of such lesser mutations there were gems like "Prelude to a Kiss," the incomparable "Blue Light," and fine arrangements of rather ordinary tunes like "Please Forgive Me." Listen, for example, in the latter, how mere timbre and sonority transform this material. Ellington's famous muted brass and the unique voices of the reed section elevate the song to a realm no other orchestra of the time could have matched—not to mention Lawrence Brown's elegant cello-like trombone lead.

"Prelude to a Kiss" is, of course, one of Ellington's most celebrated ballad songs, almost too sophisticated for wide acceptance with its sinuous melody and chromatic harmonies. Curiously, Ellington recorded it not with its lyrics, but in an instrumental version which features typically sentimental statements by Brown and Hodges.

One approaches a title such as "Battle of Swing" with some trepidation, but this late 1938 swing exercise rises above the norm—and its title. Indeed, it points brightly to the future: (1) to bop—with its zigzag figures, brisk tempo, and translation of the twelve-bar blues into a distant musical world; and (2) to Ellington's own "Cotton Tail," with its daring unison lines. Like so much Ellington, it also looks back, this time to the baroque *Concerto Grosso* concept. A solo quartet of clarinet, alto, cornet and valve trombone is pitted against the *ripieno* orchestra, the former always in harmony, the latter virtually always in unison (or octave unisons). Of the many delights this performance offers, perhaps the most striking are Rex Stewart's pixieish solos and, in the fifth chorus, Tizol's leaping Till Eulenspiegel-like figure.

Another Ellington category, the "mood" or "serenade" type of piece, is well represented in 1938 by "Stevedore's Serenade," "A Blues Serenade," "Mighty Like the Blues," and as here represented in two separate takes, the magnificent "Blue Light," which along with "Subtle Lament" (1939), "Dusk" (1940) and "Moon Mist" (1942) constitute the incomparable legacy of Ellington's all-time classic "Mood Indigo." There is nothing in all of music quite like the hushed *misterioso* atmosphere of Bigard's entrance on "Blue Light" (particularly on take one), nor had the blues changes ever been expressed so exquisitely and with such harmonic originality whilst retaining the unique trio voicings—muted trumpet and trombone with low-register clarinet—of "Mood Indigo." Brown's solo is sheer poetry, the "climax" of the piece, after which Ellington's rich piano provides the perfect resolution. If we need to have proof of the ideal wedding of content and form in jazz, we need look no further.

"Boy Meets Horn," originally entitled "Twits and Twerps"—a fair onomatopoeic rendering of those unorthodox half-valve sounds only Rex Stewart could find on a cornet—was Rex's big solo vehicle of the late 1930s and early 1940s—light fare which audiences just ate up. It could have been called "Concerto for Rex" to follow the other "concertos" of 1937.

Our year closes appropriately with "Slap Happy," which could have been called "Concerto for Carney." Brightly orchestrated to contrast with Carney's dark baritone and loping triplet figurations, spotted with unusual harmonic twists and turns, as well as a great Nanton solo—for some reason very little featured in 1938—"Slap Happy" ended the year on an upbeat, presaging the happier and even more inspired heights to which the Duke Ellington orchestra was to rise in but a few short years.

9

The Case for Ellington's Music as Living Repertory

A discussion of the controversial "whether" and "how" to perform Ellington's music as a permanent legacy. The article was written for High Fidelity, *November 1974.*

IS IT POSSIBLE—and is it right—that Ellington's music should be relegated to perpetuation solely by mechanical reproductive means? Is this remarkable musical output not to survive in live performances or perhaps only in transmutations and improvisations by others, based on the Duke's tunes?

Since Ellington's death, the factions have formed, in most cases rigidly affirming previously conceived notions. And curiously, much of the argumentation directly or indirectly opposes the perpetuation of his music as a living repertory.

The arguments run something like this. 1) Jazz is a spontaneously created, largely improvised music that cannot be recaptured for repetition. Some even say "should not." Therefore, jazz has no re-creatable repertory, as classical music does. It is constantly renewable but only in terms of improvisation, i.e., other "spontaneously created" versions of the original. It is not a music ever to be fixed.

2) Should one play Ellington's work while some of his musicians for whom the music was originally created are still alive? Indeed, his orchestra continues under his son Mercer's leadership, presumably obviating the need for others to concern themselves about the preservation of Ellington's music.

3) Since it is "impossible" to imitate the great soloists/personalities of the Ellington ensemble—Johnny Hodges, Lawrence Brown, Rex Stewart—this whole body of music is relegated to survival only in archival form, in the "museum" of recordings.

In addition there are always certain obsessively possessive jazz critics who believe that jazz is some kind of exclusive area of music belonging to them, and that treating it as repertory and thus making it available to other musicians and audiences will automatically dilute and desecrate its purity.

I cannot believe that a music as profoundly important as Ellington's (and Billy Strayhorn's) should meet such an uncertain fate. And indeed there is no reason why this music—or at least some of it—cannot continue to be played close to how it was originally conceived. The qualifying words here are "some of it" and "close."

There is, obviously, some jazz literature that could, in fact, never be re-created. One would not think of duplicating one of John Coltrane's thirty-five-minute improvisations or Eric Dolphy's amazing solos on "Stormy Weather" or indeed Hodges's "Warm Valley" performance. But Ellington's music is not limited to that kind of improvised jazz. It is well known that the Duke rejected the narrowing stigmatization of the term "jazz" for his music. And in truth much, perhaps the greater part, of his output consists of *orchestral compositions*—for a "jazz" orchestra perhaps, but an orchestra nevertheless—very often fully notated or fixed in some permanent way by himself or his musicians or both in combination. In many of these works the "improvised" solos are brief, incidental, and surprisingly "fixed" as a permanent feature of that performance. Certain "solos" were even handed down from player to player through the decades, as witness Bubber Miley's contributions from the late 1920s being played virtually the same way by his successors Cootie Williams, Ray Nance, Cat Anderson, Clark Terry, and several others. Such solos were never pure off-the-top-of-the-head improvisations to begin with. They were well-thought-out, prepared, and integrated into the total piece, and *because* this was so they were generally not tampered with by later incumbents of that chair.

This is not very far removed, if at all, from the instance of a classical composer writing a solo or a concerto, perhaps with a certain musician in mind (think of the Brahms concerto written for Joachim), which is then played by others with a slightly different style, tone, interpretation, and character.

Apart from the "solo" question in such orchestral jazz pieces, the orchestral frame is, of course, even more specifically fixed, notated, rehearsed, and played more or less the same way in each performance. It seems to me that such pieces—and Ellington created hundreds of them—are eminently suitable to performance by others if sensitively and conscientiously approached.

In answer to the second point, even when Duke was still alive a huge number of his most famous compositions were not in the band's repertory. So there were no live performances by him of such masterpieces as "Ko-Ko" or "Blue Serge" or "Azure" or "Reminiscin' in Tempo" or "Dusk." Duke undoubtedly had his reasons for not maintaining much of the old material, apart from the fact that it is simply not possible to keep over a thousand pieces in a single band's repertory. I think his reasons were mostly personal. For example, when Hodges died, virtually all the recent pieces associated with him were eliminated from the then repertory of the band, because, I think, Hodges's loss was such for Ellington that he could not bear to have anyone else play them—even if there had been someone in the band who *could* play them.

With all respect for Duke's feelings, one must say that once a composer creates a work it cannot remain the exclusive property of its creator or the person(s) for whom it was created. It belongs, in the broadest (non-copyright) sense, to the world. One simply comes back to the point that pieces as original, as perfect, as imaginative, as beautiful as Ellington's best cannot just be buried in the past. They must survive; they must be heard.

And something must be done about it before more of Ellington's music, scores and parts, disappear. Perhaps more exists than one can ascertain at this time, so soon after his death. I do know that in trying to obtain the parts for a half-dozen Ellington scores a few years ago, several days of diligent search on the part of Tom Whaley and Joe Benjamin produced nothing. Perhaps they'll turn up, but one shudders to think of the possibility that they may not.

Some will say it is enough to take some of Ellington's pieces—like "Satin Doll" or "Sophisticated Lady"—and use them as a basis for improvisations and arrangements. Unfortunately that preserves very little of Ellington. Miles Davis improvising on "Satin Doll" will come out much more Miles Davis than Duke Ellington. Furthermore most jazz musicians perform their own tunes, largely for financial reasons (like record royalties), and very few improvise on compositions by others. Beyond that, it is a fact that the majority of Ellington's music does not lend itself to that kind of improvisation. His pieces are always more than tunes, a set of changes, or a line. They are true fully thought-out compositions written for orchestra, often very complex in structure and form. Should these perish simply because they do not conform to the norm of tunes on which musicians like to blow choruses?

The remarkable fact is that a great deal of Ellington's music is *not* dependent upon performance by his own orchestra or by the Browns, Carneys, and Hodgeses. It transcends those personal qualities. It turns out that it ultimately doesn't matter whether an eight-bar "solo" by Brown, for example, in the middle of a mostly arranged composition has *exactly* Brown's tone or vibrato or slide technique. What is important is to preserve the essence and character and as much of the specifics of that "solo" as possible, because it would be difficult to conceive of anyone doing anything better in its place. Whether Brown or Ellington or both chose the notes, the result that was finally approved by Duke and performed or recorded in that form is without question the best possible realization of that musical idea or moment. *That* is what is important to preserve: the music as it was *originally* conceived, either singly by Duke or jointly by him and his musicians.

There can be little doubt that the original creative impulses and the conditions under which they occurred constitute the most complete and perfect realization. These conditions include the inspiration Ellington received from his players to create certain pieces and musical ideas for them. But it does not necessarily follow that those musical creations are limited to performance by those who first inspired them. That is obviously not true in classical music and need not be in jazz either.

In truth, Ellington's compositions are, *as* compositions, so durable that they can be played by others sensitively re-creating the original notes, pitches, rhythms, timbres, etc. But what is most astonishing is that they can, in performances by fine musicians with fine ears, not only re-create the original, but bring to it an excitement and drive that has its own validity, even though it may not be precisely the excitement that Ellington and his men got.

This is, of course, an exact parallel to classical repertory, where no two interpretations of a Brahms or Tchaikovsky symphony are the same, despite the fact that conductors and performers will be playing from the same notated parts and score. It is in that same sense that much of Ellington's music can be preserved—and *must* be. It is too important a part of our American musical legacy.

Ellington, who was always *sui generis* and conceptually ten years ahead of his contemporaries, produced an *oeuvre* that transcends the parochial views of most jazz purists. Indeed many of them did not accept or understand his musical innovations when they first appeared. It would be most inappropriate if they now would kill the growing movement toward the preservation of the jazz repertory, not only Ellington's.

10

Cecil Taylor: Two Early Recordings

A review and analysis of Cecil Taylor's first recordings, made in the late 1950s. The article first appeared in the January 1959 issue of* The Jazz Review.

THE HISTORY of harmonic-melodic developments in Western music has been—allowing for an occasional detour here and there—an almost continuous process of tonal expansion. Starting with a nucleus of harmonically fundamental tones, the triad, a concept that became crystallized during the late Middle Ages and early Renaissance developments of polyphony, the tonal boundaries were gradually expanded over the centuries to include seventh and ninth chords and all manner of chromatic deviations thereof, until early in our century the powerful hold that the "tonal center" had over musical thinking was broken, and the tonal equality and independence of the twelve tones of our chromatic scale were established. In the years just prior to this breakthrough of the tonal "sound barrier," composers such as Schoenberg, Stravinsky, Scriabin, Debussy, to name but a few, were working with the outermost extensions of tonally centered chords and melodies; and it was the increasing importance and independence that these outer extensions assumed that led to concepts of bitonality and polytonality (Stravinsky, Milhaud, etc.), and eventually pushed music across the borderline into the realm of atonality (Schoenberg, Berg, Webern, and Ives).

The history of jazz, which is taking a course virtually parallel (though in a drastically condensed form), has now reached, at least in so far as harmony and melody are concerned, a similar juncture as described above. A small minority of jazz composer-performers are

* Jazz Advance, Transition 19; Cecil Taylor Quartet at Newport, Verve MG V-8238.

working primarily with the outer reaches of tonality, and have reached that borderline where their music often spills over into areas so removed from any center of tonal gravity, that it can be thought of as "atonal." Foremost among these is Cecil Taylor, of whose work one and a half Lps are now available, with others (on Contemporary and United Artists) soon to be released.

It has been said that Cecil Taylor's music is not really atonal, and indeed he himself is quoted as saying he thinks of it "definitely as *tonal.*" Basically this seems to me to be an academic or semantic question, especially in view of the above-mentioned borderline nature of most of his playing. One can judge the work of art ultimately only with qualitative criteria. What matters in any artistic procedure or technique or system is not what it *is, but* what it can become, what it can create—a hard lesson many critics seem to have difficulty learning.

Nevertheless, since there is some confusion not only about the question of whether Taylor's music is atonal or not, but also about the whole semantics of these much bandied-about words "atonality" and "tonality," perhaps a few clarifying words should be set down before discussing the records.

Much confusion arises from the fact that the words "tonality" and "tonal" are used in two different meanings. On the one hand they are used to indicate a specific harmonic system (often wrongly equated with the diatonic system), while on the other hand they may mean in a very general way *all* intervallic relationships between *tones.* Many discussions on the subject bog down because these terms are not defined beforehand, and because the word "tonal" is often used interchangeably in both senses within even a single sentence.[1]

It is, of course, obvious that if one applies the second more general meaning, Cecil Taylor's music is tonal. His playing is even tonal very often in the other sense of the word, especially in his expositions, and in the basic fact that the bass parts in his groups so far have not ventured beyond the conventional diatonic (occasionally chromatic) walking bass-line we all know from earlier jazz. Listening carefully to his playing leaves no doubt of the fact that Taylor indeed does *think* tonally, but the result of his thinking most of the time cannot be analyzed on tonal terms (using the word now in the more specific historical sense). That is to say, the implied underlying tonal chord structure on, let's say, a blues or Ellington's "Azure," is the specific

1. It is on the basis of the second definition that some people claim that truly atonal music cannot exist, that "atonality" is a misnomer, that all music, no matter how "dissonant," is ultimately "tonal."

impetus that determines his choice of notes, especially at phrase beginnings and endings. The bulk of his improvisations, however,—and this is particularly true of the less conservative Transition Lp—is either purely atonal or is so close to the borderline between tonality and atonality, that identification and syntactical analysis via tonal centers becomes complicated out of all proportion—a kind of academic game—and meaningless.[2]

That Taylor's improvisations are in effect primarily atonal—whatever their tonal motivation may be—is indirectly attested to by certain discrepancies in the bass part, as played on these recordings by Buell Neidlinger. In the course of the proceedings, he occasionally wanders off from his "changes"—whether on purpose or not I cannot say. If one is listening objectively to the piano improvisation and the accompanying bass and drums *in their totality*—i.e., if one is not listening to the bass line by itself—these deviations in the bass seem not to matter much. They simply become absorbed in the already strongly atonal sound fabric. (In a more tonal context, such deviations would be very disturbing.) And of course, a further test—though perhaps slightly unfair to Cecil's intentions—would be to eliminate the bass part altogether. At any rate, I defy anyone to analyze the following excerpt, picked more or less at random from dozens of similar moments, in terms of tonal centrality (Ex. 1).

But then not all of Cecil Taylor's music is that advanced. It might be instructive to show those who would simply reject music such as this as senseless or "not jazz!", that it has its origins in a fairly harmless point of departure, and that what Taylor does is really quite logical and, in my opinion, imaginative and stimulating.

The opening of his own "Tune 2" on the Newport Lp, for instance, is an excellent example of the germinal ideas upon which he builds his improvised abstractions. "Tune 2" is cast in a somewhat extended form, consisting of the following schema:

A	A^1	B	C	C^1	D	B^1	C^2	D^1	B^2	D^2	A^2	
8	8	8	8	8	6	6	4	4	8	8	12	= 88 bars

(The numerals 1 and 2, qualifying the letters, are used to indicate the fact that such sections are not exact repetitions but variants of the original letter. The D sections are pedal-points using a particular prescribed rhythm, different from the rest of the piece.)

2. Discussions about whether something is atonal or tonal always remind me of current arguments about whether such and such a piece is jazz or not; they so often founder on the reef of confused semantic definition.

EXAMPLE 1 is taken from one of the later choruses of "Azure." (Measures 2 and 3 are strongly reminiscent of a certain phrase in Stravinsky's *Sacre du Printemps*).

EXAMPLE 2 is the original bridge of "Azure" upon which these particular measures in example 1 are founded.

With this as the basic compositional material, the quartet plays a short introduction, the exposition by all four, a chorus by Steve Lacy, one by Taylor, the fourth chorus divided between the two soloists, and a tagged-on 12-bar coda consisting of A^2.

As with those of many jazz soloists, Taylor's improvisations start in relative calm, close to the theme, and gradually reach a more excited and complex level as they become less tonally oriented. If our Example 1 is a typical instance of what may happen in the body of a Taylor solo, our next example (3b) is a good indication of how an improvisation might start.

Here the tonal skeleton is still quite audible (and visible), while the groundwork for further expansion is already being laid. Note the reiterated use of motive *a,* one of Taylor's favorite phrase-turns, and the use of the same material horizontally and vertically. If proof be needed that Taylor knows what he's doing and that he is not "simply faking dissonances," as some would have it, one need only point to these devices, long a mainstay of compositional techniques. One might also point to the symmetrical relationship of measures 21-22 to 23-24 (Ex. 3b), mirroring a similar relationship between 17-18 and 19-20 in the exposition (Ex. 3a). Throughout his playing, Taylor manages to retain a very close rapport between the structure of the composition and that of his improvisation, more so than Lacy does. The character of different segments of a piece like "Tune 2," for instance, are respected and employed in the improvisation, which at the same time often reveals a characterological or structural unity all its own,—as if there were two structural levels at once. At other times, the structure of a solo may momentarily take precedence over that of the composition, at which times I have the impression Cecil also cuts the tonal umbilical chord, and lets the force of the particular idea with which he is involved at the moment be the sole arbiter. At such times he reminds me of Thelonious Monk in that he—like Monk—can play passages in which the overall musical shape and direction take precedence over the actual notes; i.e., the choice of notes—though excellent—is secondary to the larger musical contours, and another, possibly equally excellent choice of notes could have rendered the same musical design.

These abilities coupled with an innate musicality give Cecil Taylor's best solos a great deal of cohesiveness. Sometimes unity is achieved by means of motivic variants and developments, sometimes by a variety of fresh ideas simply sustained at the same level of intensity.

It is in reference to the over-all continuity and sustained expressive-

EXAMPLE 3a represents the original of "Tune 2" melody with its base line and chords.

EXAMPLE 3b is the beginning of the piano improvisation thereon. (In measures 1 through 4 (not shown here), the piano is still accompanying the overlapping end of Lacy's soprano saxophone solo.)

ness however,—qualities which after all, beyond all technical considerations, determine the real validity of a musical conception—that Cecil occasionally finds himself in a dilemma. In this respect the two Lps offer a telling contrast. Where the Newport performances (possibly as a result of performer-audience contact) have an exciting, intensely felt continuity, the studio performances on Transition (except for the remarkable "Azure") suffer by and large from a lack of these qualities. One gets the impression that Taylor is sitting at the piano, *objectively* performing a function to which he was committed, and that in a rather non-participating manner he trots out a collection of ideas which, though original and varied enough, have no real artistic *raison d'etre*. One does not feel the burning necessity that what he says *had* to be said. Especially on the blues, one has the impression that Taylor lets us in on the workings of his mind, but not his soul; and he hides from us what he feels about the blues. But as I say, the Newport record and several excellent live performances of Taylor that I have heard, indicate that such moments are in the minority.

Yet it is a point worth discussing because it relates to the whole question of atonal improvisation or, for that matter, atonal jazz composition. The performer is caught between two cross-fires, as it were. He wants to free himself from the conventional tonal strictures set down by the bass (or at least implied by the chord patterns upon which he is supposed to be improvising), but at the same time, as long as the bass and drums participate in the accepted conventional manner, the improviser will constantly feel the gravitational pull of both their tonal and rhythmic weight. One can hear this on the Transition Lp, especially on "Charge 'Em Blues." Time and time again Taylor is pulled back from his intentions by the conventions of tonality, of phrase lengths and the "beat." The rhythm section, pushing relentlessly forward in its duty as supplier of the beat, sets a perfect trap for the improviser. It propels him onward while he haphazardly clutches at ideas that may come to him—some good, some commonplace. The relentlessness of the accompaniment can easily push him into mechanical and rhetorical solutions, unless he can free himself from their influence.[3]

Thus in "Charge 'em Blues" Taylor starts an idea, spins it out a

3. I realize that these problems exist in all ordinary jazz improvisation, but the problem is even more acute when the improviser has the added burden of making a less familiar atonal context intelligible.

little by repeating it[4] or using slightly varied imitations, then breaks it off and starts another idea. It's almost like listening to a solo consisting entirely of "fours." To sustain this enormous variety of ideas in this manner is, of course, extremely difficult, and probably impossible in pure improvisation. To help himself, Cecil had several musical ideas in readiness (others might call them clichés, although let it be said that they are at least his own) which he could throw in whenever necessary. Example 4 is typical; in various guises it appears at least a dozen times on the Transition Lp.

I called this situation a dilemma, and I really do not mean to criticize Cecil for failing to solve it. For no doubt the problem is a difficult one. If the soloist is strong enough to resist this pull of the bass and drums, and can soar in relative autonomy above the accompaniment, the question promptly arises: well, why not eliminate the bass and drums? But that in turn immediately produces the next question, which is: can this bass-less, drum-less, atonal wonder still be considered jazz?[5]

However, my impression is that Cecil Taylor *is* concerned about playing jazz; and to his everlasting credit, on the Newport Lp—made almost a year after the Transition recording—Taylor walks the tightrope between the two cross-fires with razor-sharp accuracy, and with all the emotional, swinging intensity associated with good jazz. The segmentation which marred the earlier recording is no longer in evidence. The ideas—again presented with a seemingly endless variety—all relate logically and smoothly to each other, so that one hears a musical edifice, made up of many contrasting elements, but all coalesced into an expressive entity.

I like, aside from Cecil's solos, his accompaniments behind Lacy, and his ability to relax into a groovy, slightly-behind-the-beat swing. I noticed too, that in "Tune 2" the entire group loosened up to swing

4. The question of whether Taylor's playing is to be considered atonal or not comes up again in connection with the concept of repetition. Nearly a half a century of atonal composing has convinced most composers that forms of symmetry, including repetition—especially *immediate repetition*—are out of place in atonality, because they have lost their functional ties with the symmetrical tonic-dominant relationship that governed diatonic music.

5. Other questions that arise in this connection are: Is it logical at all to mix an atonal improvisation with a tonal bass line? Does it make sense to pour a highly complex idea into a conventional rhythmic mold? Do not such discrepancies ultimately detract from the stature and validity of such a concept?

EXAMPLE 4

every time the last twelve bars (starting with the A minor chord) appeared. It seems that the pent-up tension of the preceding eight-bar pedal-point (on E) found release in the resolution to A. It worked all four times.

One thing the group has yet to learn is not to rush, especially on medium tempos. "Tune 2," for instance, picks up considerable speed before the exposition is sixteen bars old! But at least this is a sign of life and energy.

Buell is a continually improving bass player. His tone is rich and full (meaty is the word), but it does not quite fit in character with the harder, more acrid quality of Cecil's playing. As a matter of fact, a player with more bite could add to the quartet-ness of the group. One thinks of Mingus, for instance, as ideal. Buell's occasional erring, as on the end of "Azure," can be forgiven, considering the fact that the bass player's lot in a group like this is a very lonely one.

This brings us to the point of whether it matters ultimately if we call it jazz or not. Again, should it not be sufficient to ask how *good* it is, rather than *what* it is?

Dennis Charles, the drummer, is a sympathetic accompanist and often enters into interesting, though by no means startling, musical exchanges with his leader.

Steve Lacy I respect and admire for his consistent astute work, but I wish he would avoid using such a choked *tarrogato*-like sound, especially in the low register. It's improbable, but I suppose it could be argued that this is the sound he or the group wants. But I find it terribly distracting—and that, I don't think, was its intention.

In discussions on music such as this one often hears the expression "twelve-tone jazz" bandied about. It seems like such a nice, new, catchy word. I should like to make it explicitly clear that the term "twelve-tone" in connection with jazz is only applicable to written or composed jazz. "Twelve-tone improvisation" *does not and can not exist*. The procedures of twelve-tone or serial composition are of considerable complexity, and, if they were applied to pure improvisation, the improviser would have to have a memory and calculating ability greater than Univac's. Even if it were possible, I doubt if it were desirable. This is not to say that, since they cannot be used for intuitive improvisation, methods of serial composition are entirely mathematical or inhuman; it simply means that their use entails more time than is available in extemporization.

11

Ornette Coleman

Schuller's entry on the controversial saxophonist for the New Grove Dictionary of American Music.

COLEMAN, ORNETTE (b. Fort Worth, 19 March 1930). Black American jazz saxophonist and composer (and in later years trumpeter and violinist). He began playing the alto saxophone at fourteen, developing in ensuing years an idiom predominantly influenced by Charlie Parker. Early professional work, however, with a variety of southwestern "rhythm and blues" and carnival bands seems to have been in more traditional idioms. In 1948 Coleman moved to New Orleans, working mostly at non-musical jobs. By 1950 he had returned to Fort Worth, subsequently going on to Los Angeles with the "Pee Wee" Crayton "rhythm and blues" band. Wherever he tried to introduce some of his more personal and innovative ideas, he met with hostility, both among audiences and musicians. Working as an elevator operator in Los Angeles, he studied (on his own) harmony and theory textbooks, as well as saxophone studies. It is in those years that Coleman evolved a radically new concept and style, seemingly out of a combination of musical intuitions born of southwestern country blues and folk forms, and misreading (in part)—or highly personal interpretations—of those theoretical texts.

Eking out his income by working sporadically in some of downtown Los Angeles's more obscure and menial entertainment clubs, Coleman eventually came to the attention of bassist Red Mitchell, who in turn alerted Percy Heath of the Modern Jazz Quartet. Coleman's first studio recording (for Contemporary, in 1958) reveals that his style and sound were in essence fully formed. At pianist John Lewis's instigation, Coleman (and his trumpet partner Don Cherry) attended the Lenox School of Jazz in Massachusetts in 1959. There followed engagements at the Five Spot in New York as well as a series of recordings in 1959 (for Atlantic) entitled *The Shape of Jazz To Come,* including his compositions "Lonely Woman" and "Congeniality," and *Change of the Century* with "Ramblin'" and "Free." These performances, which brought Coleman to worldwide (and controversial) attention, revealed a style freed from most of the conventions of "modern jazz." Highly flexible in tonality, rhythmic continuity, and form in the small sense, Coleman's improvisations liberated the jazz solo both from an adherence to predetermined harmonic "changes" and a subservience to melodic (thematic) variation. Traditional chorus and phrase structuring were also abandoned, at the same time reinterpreting jazz rhythm, beat, and swing along free non-symmetrical lines. Although Coleman's playing seemed to many to be incoherent and "atonal," it was (and is to this day) essentially modal in concept, well rooted in older, simpler black folk-idioms.

Coleman's music cannot be understood solely in terms of the generally prevailing concept (since the late 1920s) of jazz as a virtuoso soloist's form of expression, for it is intrinsically a spontaneous polyphonic ensemble art. It is founded–here again it reveals its older roots–on a consistent use of collective interplay at the most intimate and intricate levels of instantaneous mutual reaction. Hence the music's extraordinary unpredictableness, freedom, and flexibility. The 1960 recording *Free Jazz* (on Atlantic) for double jazz quartet, a thirty-seven-minute sustained collective improvisation, was a remarkably prophetic display of the new-yet-old concept, and was undoubtedly the single most important influence on avant-garde jazz of the ensuing decade.

Another 1960 recording *Jazz Abstractions* (also on Atlantic) revealed Coleman in a variety of more structured compositional frameworks, particularly in a serial work, "Abstraction" by Gunther Schuller, for alto saxophone, string quartet, two basses, guitar, and percussion.

In 1962 Coleman retired temporarily from public performing, primarily to study trumpet and violin (again self-taught). When he returned to public life in 1965, his unorthodox treatment of these latter instruments provoked even more controversy and led to numerous denunciations of his work by a number of influential American jazz musicians, including Miles Davis, Benny Carter, Charles Mingus. However, Coleman was well received in Europe on his first tour there in 1965, giving a major impetus to the burgeoning European jazz avant-garde movement of the time.

In the mid- and late 60s, Coleman became interested in extended and more "formal" through-composed work for larger ensembles, producing among others *Forms and Sounds* for woodwind quintet (1965) and *Skies of America,* a 21-movement suite for symphony orchestra (1972).

By the early 1970s Coleman's influence waned considerably, while John Coltrane's dominance of saxophone styles spread correspondingly. As Coleman turned increasingly to more abstract and mechanical compositional techniques, as exemplified in *Skies of America,* his playing lost some of its earlier emotional intensity and rhythmic vitality. But a visit to Morocco in 1973 and the gradual (especially rhythmic) influence of certain popular rock, funk, and fusion styles seem to have revitalized his ensemble performances, a direction clearly discernible in Coleman's powerful 1981 electric band, Prime Time.

While a full assessment of Coleman's work and influence is at this point perhaps premature, certain characteristics seem to be consistently present in and essential to his style. In the context of Coleman's polyphonic and tonally free ensemble performances, his own improvisations may at first sound "atonal." But closer listening reveals a basically modal melodic approach, which preserves much of the simplicity of older black folk-idioms. It contains in particular the raw "cry of the blues." A musical humanist and philosopher at heart, Coleman's wailing saxophone sound (produced in his early years on a plastic saxophone) is never far removed from the plaintive human voice of Afro-American musical folklore. This essentially lyric approach, best heard on "Lonely Woman" (1959) and "Sex Spy" (1977), is linked to Coleman's "horizontal" concept of improvisation, a tendency explored earlier by such players as Lester Young and Miles Davis (in his post-be-bop modal style). Released from a strict adherence to harmonic functions and conventional form and phrase patterns, Coleman's solos are intrinsically linear, evolving in a free-association, sometimes fragmented musical discourse.

His faster tempo improvisations are marked by flurries of notes; or gliding, swooping, at times bursting phrases, played with great intensity and conviction. Occasionally his work seems to be burdened by the over-use of sequential patterning. But it is finally the strength of conviction of his playing, especially when aided and abetted by like-minded colleagues (Don Cherry, bassist Charlie Haden, drummer Billy Higgins), that produces a sense of the inevitable in Coleman's art.

Technically Coleman plays as much "from his fingers" as from his ears, an approach frequently resulting in non-tempered intonations and unequal tone colorings. (These effects are even more noticeable in his less than convincing trumpet and violin playing, although even on these instruments Coleman can sometimes produce by sheer instinct and musical energy the most compelling improvisations.)

Coleman's style has changed little since the early 1960s. In its freely modal stance, it can function successfully in a wide variety of contexts. Whether working in native Moroccan musical traditions or in "atonal" classically oriented works or, indeed, in rock/funk-influenced idioms, Coleman's playing in both sound and substance seems to be capable at once of dominating its surroundings and assimilating into them.

In recent years Coleman has espoused a theory which he calls "harmolodic." It is apparently based on the untransposed performing in varied clefs and "keys" of the same musical materials (lines, themes, melodies), thus producing a simplistic organum-like "polyphony," primarily in parallel unrelieved motion. It is not clear, however, how this theory functions in Coleman's own improvisatory style.

Coleman is also noted for his obscure, often inherently contradictory verbal epigrams. Some observers see in them the "philosophical" analogues to his musical theories and concepts. In an interesting corollary, his notations of his own compositions—several hundred are known to exist—are imprecise, gestural, and in a sense graphic, leaving the interpreter free to give his ideas individual and differing realizations.

While it may be impossible as yet to assign a specific and all-pervasive Coleman influence in jazz (in the sense that one can do so with Coleman Hawkins, Lester Young, Charlie Parker, John Coltrane), it is nonetheless clear that Coleman opened up unprecedented musical vistas for jazz, the wider implications of which have not yet been fully explored (least of all by his many lesser imitators).

12

Ornette Coleman's Compositions

The Foreword to a collection of ten compositions by Coleman (MJQ 6), published by MJQ Music Inc. in 1961. The publication consisted of Schuller's notated transcriptions from recordings of two of Coleman's early Lps, made for the Atlantic label (1317 and 1327).

THIS COLLECTION of compositions by Ornette Coleman is probably in certain respects one of the most unusual music publications ever undertaken. Ordinarily, publication takes place once a composer has submitted a manuscript to the publisher. Where this has not been the case, as in the case of folksong collections, for example, the music has usually been of such a simple nature (or has been so simplified by the editor) that difficulties of the kind encountered in this edition never arose.

Mr. Coleman, a controversial alto saxophonist and composer, through an undoubtedly unique set of circumstances was "spared" conventional musical education. Despite the fact that he had played the saxophone some fifteen years before he made his first recording, Mr. Coleman never learned to read or write conventional musical notation correctly. In the history of human civilization this is, of course, not unusual; however, in the context of our Western civilization, a musician's total immunity to the notational aspect of music must be considered somewhat of an exception. And while, in the early days of jazz, non-reading improvising musicians were the rule rather than the exception, the reverse is true today, and most jazz musician's can read at least moderately well. Not so Mr. Coleman. Lest this be construed as criticism of his abilities, we wish to assure the reader that, were this the case, this publication would never have been undertaken. On the contrary, we believe it is precisely because Mr. Coleman was not "handicapped" by conventional music education that he has been able to make his unique contribution to contemporary music.

The specific problems of notating these compositions arise from the fact that the "leadsheets" submitted by Mr. Coleman, written in a highly "personal" notation, rarely coincided rhythmically with his own performances of these works. The editor, therefore, was forced to transcribe them from Mr. Coleman's recordings. A further complication arose at this point, since in those instances, where the rhythmic notation was *not* unmistakably clear, Mr. Coleman was unable to verify one way or another the editor's particular choices.

It must therefore be emphasized that while every effort at accuracy has been made, the editor cannot claim to have solved every notational problem unequivocally.

Notation is in many ways based on arbitrary decisions by the composer (for example whether to notate a piece at ♩ = 126 or 𝅗𝅥 = 126). Similarly, in this edition, lacking of necessity the authority of the composer's personal choice, arbitrary decisions had to be made by the editor. However, we believe they are limited to instances where the recorded performance was itself not definitive enough to arrive at an unequivocal choice. The paradox of the situation is that a composer's performance of his own music is unquestionably a more direct and accurate source of his intentions than his notation of the same music could ever be. However, for the publisher, who obviously cannot exist without notation, the recording by itself—no matter how authoritative—lacks the composer's written corroboration of his performance.

To complicate matters more, there are certain variables in the recorded performances of Mr. Coleman's works with respect to tempos, meters and chord progressions, not only within a given piece, but from performance to performance, so that the editor was unable to check his notational choices against live performances.

In Mr. Coleman's world, where freedom and constant variation are the main guiding principles, it is not at all unusual that an improvised phrase, which on its first appearance consists of six bars, turns out to be eight the next time and perhaps nine still later. The chord progressions (or "changes") upon which the improvisations are based are often only loosely related to the composition, at other times relate only to *part* of the written section, and in any case are not always strictly adhered to. (See discussion of "Congeniality" below.) In "Lonely Woman" another type of performance variability occurs: in this piece only the drums keep a strict tempo, while alto and trumpet (in an "improvised" unison) and the bass play not only at different tempos, but within these tempos quite freely (*rubato*).

These structural and harmonic liberties have led some observers to conclude that Mr. Coleman's music is chaotic and formless. But even a perfunctory glance at the ten compositions in this book will indicate that this is not the case. In fact, the pyramid-like form of "Focus on Sanity," with its related tempo levels; the ingenious meter relationships of "Una Muy Bonita"; the simple yet original formal designs of all the other pieces, would indicate that Mr. Coleman's music is anything but chaotic. The originality of these compositions is all the more startling when one remembers that they are *not* the product of book learning or conventional musical education, but instead intuitive creations whose genuineness is for this reason alone unassailable.

It has also been said that Ornette Coleman is a fine composer but a poor improviser. The inference is that, because he rarely adheres to conventional chord or phrase patterns, he is incapable of doing so, and that therefore his improvisations are "fraudulent" or at best "disorganized" and "meaningless." For this reason the editor has included one of Mr. Coleman's improvisations ("Congeniality") to place before the unprejudiced musician an example of this alleged "incoherence."

A comparison of the written-down improvisation with the actual recording will show that Mr. Coleman is quite capable of making coherent statements which, though their external continuity may be fragmented, have an unmistakable phrase-to-phrase logic. Moreover, there are interesting internal relationships within the entire solo which offer concrete evidence that Mr. Coleman is not simply "noodling" or lost in an a-harmonic maze. For example, motivic fragment *a* in the first measure is immediately re-used and varied (bars 3 and 5). Another fragment *c* appears at four different places in the solo. Motive *d* occurs three times, and is probably half-consciously derived from the sequence of notes which first appears in the eighth measure of the theme. A simple but interesting device (probably first used extensively by Coleman Hawkins) is employed between letter *C* and *D*, when fragment *f*, first used to conclude motive *e*, appears a second time tacked on to a transposed variant of *e*, and then a third time to launch a new phrase. The use of the rhythmic figure *j* in conjunction with the same drop of a major third three times would indicate some organizational process. There are also numerous examples of sequential or quasi-sequential treatment (motives *b*, *e*, *g* and *h*).

Aside from such melodic relationships, there are many indications in Mr. Coleman's harmonic patterns that "he knows what he's doing." He is basically a modal player, a point which underscores how far back his musical roots reach. An analysis of the harmonic progression seemingly implied by his solo in "Congeniality" indicates clearly that Mr. Coleman does not veer much from the basic key of B-flat,—and then most often only to the next step of C minor. Excursions into other keys (like D-flat and B-natural) are rare and momentary, and all follow more or less the same pattern. The sequence D-flat, B-natural, C-minor to B-flat, for example, occurs three times (see measures 10, 56 and 160),—surely not mere accident. Analysis of other improvisations by Mr. Coleman indicate that he, like many other jazz artists, past and present, has found a way of making the tonic and *one* other step in the key serve for most of his solo: in this case the adherence to B-flat and C-minor, a combination (I and II) he seems to prefer above all others, probably because II includes most of the important notes of the dominant (V) and sub-dominant (IV). It is also obvious from the long stretches of B-flat tonic, mostly centering around the beginnings of what appear to be larger phrase structures, that Mr. Coleman is fully aware of his place in the over-all formal design at any given moment. That this structure need not necessarily consist of eight-bar units, or indeed of any particular unit length, is one of Mr. Coleman's fundamental departures from previous practices. In this connection his own statement made in 1958 is revealing: "I would prefer it if musicians would play my tunes with different changes as they take a new chorus, so that there'd be all the more variety in the performance."

Since Mr. Coleman's quartet does not employ a piano, the bassist is free to build long melodic lines which are based on a purely intuitive, reflexive reaction to Mr. Coleman's playing, who in turn responds in kind to the bass, so that a kind of continuous contrapuntal exchange is established. This explains why Charles Haden's bass lines do not always match bar for bar Mr. Coleman's harmonic patterns. Rather than mesh perfectly, they hover about each other, leaving both players free to strike out on new paths at the right moment, and pull the other one with him. This process is really the essence of collective improvisation, and is seen in a new light in the work of Mr. Coleman's quartet.

Because the improvised bass part under the alto solo in "Congeniality" does not function merely in terms of harmonic roots, but moves rather as a melodically free agent, the editor felt that its inclusion would serve no purpose in explaining Mr. Coleman's improvisation, and might, therefore, confuse those who are used to viewing music only harmonically and vertically.

From our harmonic analysis (based on its horizontal continuity) it can be seen that Mr. Coleman's work, which has often been characterized as "atonal," is in the strictest sense not that at all—certainly not in its orientation. "Free" in this context seems to be too readily confused with "atonal." What does happen, however, is that both alto and bass may move at a particular moment into divergent keys, thus giving the impression momentarily of a pan-tonal or atonal texture.

There remains to be discussed the curious 3/4 and 5/4 meters before and after letter F. In bar 121 motive *d* is altered to produce an unexpected polymetric pattern with the bass. At the conclusion of this phrase (bar 127), whether by accident or intent, Mr. Coleman starts a new phrase which is one beat off from the rhythm section. Now it is easy to assume that Mr. Coleman erred in his entrance. Yet this kind of polymetric displacement is the heart and soul of one of the primary antecedents of jazz, namely native African music, and is certainly not unknown in jazz improvisation. It is conceivable then that Mr. Coleman instinctively and deliberately made this choice, especially in as much as he likes to "turn phrases around on a different beat, thereby raising the freedom of my playing."

In the ensuing five bars Mr. Coleman heard that his rhythm section was beginning to grow wobbly under him. He accordingly stretched the next bar into a 5/4 bar, thus making up for the previously missing beat. In the meantime bass and drums, however, had begun to switch their beat, which was accomplished by bar 136. Mr. Coleman, now back in his original beat pattern, once more found himself at odds with the rhythm section. It took another 3/4 bar (142) to right things once again. This curious bit of metric interchange may have been accidental; one's conclusion can be no more than an *interpretation,* and therefore not really conclusive. However, it is this kind of rhythmic freedom which Mr. Coleman and other young players are striving for, and this example may well be an authentic forecast of things to come.

All chord symbols are additions of the editor and are to be taken merely as suggestions, in an attempt to give this collection greater practical value. Occasionally Mr. Haden's bass lines have been notated in small print, and in view of the above-outlined approach of the players, these bass lines do not always conform to indicated chord symbols. Interested players are, of course, at liberty to create their own improvised lines. Since this collection contains what are in effect condensed scores, alto and trumpet are both notated at actual sounding pitch. Parts would have to be transposed.

In general, it should be remembered that in view of the fact that jazz is largely an improvised music, in which the personal interpretation takes precedence over the composition, these scores were intended to serve only as a moderately detailed outline to help the student, music lover and aficionado.

13

Sonny Rollins and the Challenge of Thematic Improvisation

An analysis of Sonny Rollins's recorded performance of Blue 7 *(Prestige 7079), written for the first issue of* The Jazz Review *(November 1958).*

SINCE THE DAYS when pure collective improvisation gave way to the improvised solo, jazz improvisation has traveled a long road of development. The forward strides that characterized each particular link in this evolution were instigated by the titans of jazz history of the last forty-odd years: Louis Armstrong; Coleman Hawkins; Lester Young; Charlie Parker and Dizzy Gillespie; Miles Davis; collectively the MJQ under John Lewis's aegis; and some others in varying but lesser degrees. Today we have reached another juncture in the constantly unfolding evolution of improvisation, and the central figure of this present renewal is Sonny Rollins.

Each of the above jazz greats brought to improvisation a particular ingredient it did not possess before, and with Rollins thematic and structural unity have at last achieved the importance in *pure* improvisation that elements such as swing, melodic conception and originality of expression have already enjoyed for many years.

Improvisatory procedures can be divided roughly into two broad and sometimes overlapping categories which have been called *paraphrase* and *chorus* improvisation. The former consists mostly of an embellishment or ornamentation technique, while the latter suggests that the soloist has departed completely from a given theme or melody and is improvising freely on nothing but a chord structure. (It is interesting to note that this separation in improvisational techniques existed also in classical music in the sixteenth to eighteenth centuries, when composers and performers differentiated between ornamentation (*elaboratio*) and free variation (*inventio*).) Most improvisation in the modern jazz era belongs to this second category, and it is with developments in this area that this article shall concern itself.

In short, jazz improvisation became through the years a more or less unfettered, melodic-rhythmic extemporaneous composing process in which the sole organizing determinant was the underlying chord pattern. In this respect it is important to note that what we all at times loosely call "variation" is in the strictest sense no variation at all, since it does not proceed from the basis of varying a given thematic material but simply reflects a player's ruminations on an *unvarying* chord progression. As André Hodeir put it in his book *Jazz: Its Evolution and Essence,* "Freed from all melodic and structural obligation, the chorus improvisation is a simple emanation inspired by a given harmonic sequence."

Simple or not, this kind of extemporization has led to a critical situation: to a very great extent, improvised solos—even those that are in all other respects very imaginative—have suffered from a general lack of over-all cohesiveness and direction—the lack of a unifying force. There are exceptions to this, of course. Some of the great solos of the past (Armstrong's "Muggles," Hawkins's "Body and Soul" (second chorus), Parker's "Ko-Ko," etc.) have held together as perfect compositions by virtue of the improviser's genial intuitive talents. (Genius does not *necessarily* need organization, especially in a strict academic sense, since it makes its own laws and sets its own standards, thereby creating its own kind of organization.) But such successful exceptions have only served to emphasize the relative failure of less inspired improvisations. These have been the victims of one or perhaps all of the following symptoms: (1) The average improvisation is mostly a stringing together of unrelated ideas; (2) Because of the *independently* spontaneous character of most improvisation, a series of solos by different players within a single piece have very little chance of bearing any relation to each other (as a matter of fact, the stronger the individual personality of each player, the less uniformity the total piece is likely to achieve); (3) In those cases where composing (or arranging) is involved, the body of interspersed solos generally has no relation to these nonimprovised sections; (4) Otherwise interesting solos are often marred by a sudden quotation from some completely irrelevant material.

I have already said that this is not altogether deplorable (I wish to emphasize this), and we have seen that it is possible to create pure improvisations which are meaningful realizations of a well-sustained over-all feeling. Indeed, the majority of players are perhaps not temperamentally or intellectually suited to do more than that. In any case, there is now a tendency among a number of jazz musicians to bring thematic (or motivic) and structural unity into improvisation. Some do this by combining composition and improvisation, for instance the Modern Jazz Quartet and the Giuffre Three; others, like Sonny Rollins, prefer to work solely by means of extemporization.

Several of the latter's recordings offer remarkable instances of this approach. The most important and perhaps most accessible of these is his "Blue 7" (Prestige LP 7079). It is at the same time a striking example of how *two* great soloists (Sonny and Max Roach) can integrate their improvisations into a unified entity.

I realize fully that music is meant to be listened to, and that words are not adequate in describing a piece of music. However, since laymen, and even many musicians, are perhaps more interested in knowing exactly how such structural solos are achieved than in blindly accepting at face value remarks such as those above, I shall try to go into some detail and with the help of short musical examples give an account of the ideational thread running through Rollins's improvisation that makes this particular recording so distinguished and satisfying.

Doug Watkins starts with a restrained walking bass-line and is soon joined by Max Roach, quietly and simply keeping time. The noncommital character of this introductory setting gives no hint of the striking theme with which Rollins is about to enter. It is made up of three primary notes: D, A flat, and E.[1] (Ex. 1). The chord progression underlying the entire piece is that of the blues in the key of B flat. The pri-

EXAMPLE I

1. The notes C, D flat, and A in bar 5 are simply a transposition of motive *a* to accommodate the change to E flat in that measure, and all other notes are nonessential alterations and passing tones.

mary notes of the theme (D, A flat, E) which, taken by themselves, make up the essential notes of an E-seventh chord thus reveal themselves as performing a double function: the D is the third of B flat and at the same time the seventh of E; the A flat is the seventh of B flat and also (enharmonically as G sharp) the third of E; the E is the flatted fifth of B flat and the tonic of E. The result is that the three tones create a bitonal[2] complex of notes in which the "blue notes" predominate.

At the same time, speaking strictly melodically, the intervals D to A flat (tritone) and A flat to E (major third) are among the most beautiful and most potent intervals in the Western musical scale. (That Rollins, whose music I find both beautiful and potent, chose these intervals could be interpreted as an unconscious expression of affinity for these attributes, but this brings us into the realm of the psychological and subconscious nature of inspiration and thus quite beyond the intent of this article.)[3]

This theme then—with its bitonal implications (purposely kept pure and free by the omission of the piano), with its melodic line in which the number and choice of notes is kept at an almost rock-bottom minimum, with its rhythmic simplicity and segmentation—is the fountainhead from which issues most of what is to follow. Rollins simply extends and develops all that the theme implies.

As an adjunct to this twelve-bar theme, Rollins adds three bars which in the course of the improvisation undergo considerable treatment. This phrase is made up of two motives. It appears in the twelfth to fourteenth bars of Rollins's solo, (Ex. 2) and at first seems gratu-

EXAMPLE 2

2. Bitonality implies the simultaneous presence of two tonal centers or keys. This particular combination of keys (E and B flat—a tritone relationship), although used occasionally by earlier composers, notably Franz Liszt in his *Malediction Concerto,* did not become prominent as a distinct musical device until Stravinsky's famous *"Petrushka chord"* (F sharp and C) in 1911.

3. It should also be pointed out in passing that "Blue 7" does not represent Rollins's first encounter with these particular harmonic-melodic tendencies. He tackled them almost a year earlier in "Vierd Blues" (Prestige LP 7044, Miles Davis Collector's Items). As a matter of fact, the numerous similarities between Rollins's solos on "Blue 7" and "Vierd Blues" are so striking that the earlier one must be considered a study or forerunner of the other. Both, however, are strongly influenced, I believe, by Thelonious Monk's explorations in this area in the late forties, especially such pieces as "Misterioso" (Blue Note LP 1511, Thelonious Monk, Vol. 1).

itous. But when eight choruses later (eight counting only Rollins's solos) it suddenly reappears transposed, and still further on in Rollins's eleventh and thirteenth choruses (the latter about ten minutes after the original statement of the phrase) Rollins gives it further vigorous treatment, it becomes apparent that it was not at all gratuitous or a mere chance result, but part of an over-all plan.

A close analysis of Rollins's three solos on "Blue 7" reveals many subtle relationships to the main theme and its three-bar sequel. The original segmentation is preserved throughout. Rollins's phrases are mostly short, and extended rests (generally from three to five beats) separate all the phrases—an excellent example of how well-timed silence can become a part of a musical phrase. There are intermittent allusions to the motive fragments of his opening statement. At one point he introduces new material, which, however, is also varied and developed in the ensuing improvisation. This occurs four bars *before* Max Roach's extended solo. A partial repetition of these bars *after* Max has finished serves to build a kind of frame around the drum solo.

In this, Rollins's second full solo, thematic variation becomes more continuous than in his first time around. After a brief restatement of part of the original theme, Rollins gradually evolves a short sixteenth-note run (Ex. 3) which is based on our Ex. 1, motive *a*. He reworks

EXAMPLE 3

this motive at half the rhythmic value, a musical device called diminution. It also provides a good example of how a phrase upon repetition can be shifted to different beats of the measure thus showing the phrase always in a new light. In this case Rollins plays the run six times: as is shown in Ex. 3 the phrase starts once on the third beat, once on the second, once on the fourth, and three times on the first beat.[4]

Another device Rollins uses is the combining and overlapping of two motives. In his eighth chorus, Rollins, after reiterating Ex. 2, motive *a*, continues with motive *b*, but without notice suddenly converts it into another short motive (Ex. 4) originally stated in the second chorus. (In Ex. 5 the small cue-sized notes indicate where Rollins would have gone had he been satisfied with an exact transposition of the phrase; the large notes show what he did play.)

EXAMPLE 4

EXAMPLE 5

But the crowning achievement of Rollins's solo is his eleventh, twelfth, and thirteenth choruses (Ex. 6) in which, out of twenty-eight measures, all but six are directly derived from the opening and two further measures are related to the four-bar section introducing Max's drum solo. Such structural cohesiveness—without sacrificing expressiveness and rhythmic drive or swing—one has come to expect from the composer who spends days or weeks writing a given passage. It is another matter to achieve this in an on-the-spur-of-the-moment extemporization.

The final Rollins touch occurs in the last twelve bars in which the theme, already reduced to an almost bare-bones minimum, is drained of all excess notes, and the rests in the original are filled out by long held notes. The result is pure melodic essence (Ex. 7). What more perfect way to end and sum up all that came before!

4. It is also apparent that Rollins had some fingering problems with the passage, and his original impulse in repeating it seems to have been to iron these out. However, after six attempts to clean up the phrase, Rollins capitulates and goes on to the next idea. Incidentally, he has experimented with this particular phrase in a number of pieces and it threatens to become a cliché with him.

EXAMPLE 6: *a* is derived from our Ex. 2, motive *a; b* from Ex. 2, motive *b; c* from Ex. 1; *d* from Ex. 4; *f* from Ex. 1, motive *a;* and g comes from the same, using only the last two notes of motive *a; e* is derived from the new material used in the "frame" passage around Max's solo.

Bar 26 in this example is an approximation; Rollins delays each repetition by a fraction of a beat in such a way that it cannot be notated exactly.

EXAMPLE 7

This then is an example of a real variation technique. The improvisation is based not only on a harmonic sequence but on melodic/motivic ideas as well.[5] It should also be pointed out that Rollins differs from lesser soloists who are theme-conscious to a certain extent, but who in practice do not rise above the level of exact repetition when the chords permit, and when they don't, mere sequential treatment. Sequences are often an easy way out for the improviser, but easily become boring to the listener. (In fact, in baroque music, one of the prime functions of embellishment techniques was to camouflage harmonically sequential progressions.) In this respect Rollins is masterful since in such cases he almost always avoids the obvious and finds some imaginative way out, a quality he has in common with other great soloists of the past, e.g., Prez, Parker, etc.

On an equally high level of structural cohesiveness is Max Roach's aforementioned solo. It is built entirely on two clearly discernible ideas: (1) a triplet figure which goes through a number of permutations in both fast and slow triplets, and (2) a roll on the snare drum. The ingenuity with which he alternates between these two ideas gives not only an indication of the capacity of Max Roach as a thinking musician, but also shows again that exciting drum solos need not be just an *un*thinking burst of energy—they can be interesting and meaningful compositions. Behind Rollins Max is a fine accompanist, occasionally brilliantly complementing Sonny's work, for example eleven bars after his drum solo, when he returns with a three-bar run of triplets followed a second later by a roll on the snare drum—the basic material of his solo used in an accompanimental capacity.[6]

5. In this Rollins has only a handful of predecessors, notably Jelly Roll Morton, Earl Hines, Fats Waller, and Thelonious Monk, aside from the already mentioned Lewis and Giuffre.

6. A similarly captivating instance of solo thematic material being used for accompanimental purposes occurs in the first four bars of John Lewis's background to Milt Jackson's solo in "Django" (Prestige LP 7057).

Such methods of musical procedure as employed here by Sonny and Max are symptomatic of the growing concern by an increasing number of jazz musicians for a certain degree of intellectuality. Needless to say, intellectualism here does not mean a cold mathematical or unemotional approach. It does mean, as by definition, the power of reason and comprehension as distinguished from *purely* intuitive emotional outpouring. Of course, purists or anti-intellectualists (by no means do I wish to *equate* purists with anti-intellectuals, however) deplore the inroads made into jazz by intellectual processes. Even the rather reasonable requisite of technical proficiency is found to be suspect in some quarters. Yet the entire history of the arts shows that intellectual enlightenment goes more or less hand in hand with emotional enrichment, or vice versa. Indeed the great masterpieces of art—any art—are those in which emotional *and* intellectual qualities are well balanced and completely integrated—in Mozart, Shakespeare, Rembrandt. . . .

Jazz too, evolving from humble beginnings that were sometimes hardly more than sociological manifestations of a particular American milieu, has developed as an art form that not only possesses a unique capacity for individual and collective expression, but in the process of maturing has gradually acquired certain intellectual properties. Its strength has been such that it has attracted interest in all strata of intellectual and creative activity. It is natural and inevitable that, in this ever broadening process, jazz will attract the hearts and minds of all manner of people with all manner of predilections and temperaments—even those who will want to bring to jazz a roughly five-hundred-year-old musical idea, the notion of thematic and structural unity.

And indeed I can think of no better and more irrefutable proof of the fact that discipline and thought do not necessarily result in cold or unswinging music than a typical Rollins performance. No one swings more (hard or gentle) and is more passionate in his musical expression than Sonny Rollins. It ultimately boils down to how much talent an artist has; the greater the demands of his art—both emotionally and intellectually—the greater the talent necessary.

A close look at a Rollins solo also reveals other unusual facets of his style: his harmonic language for instance. Considering the astounding richness of his musical thinking, it comes as a surprise to realize that his chord-repertoire does not exceed the normal eleventh or thirteenth chord and the flatted-fifth chords. He does not seem to require more and one never feels any harmonic paucity, because within this limited language Rollins is apt to use only the choicest notes, both harmonically and melodically, as witness the theme of "Blue 7." Another characteristic of Rollins's style is a penchant for anticipating the harmony of a next measure by one or two beats. This is a dangerous practice, since in the hands of a lesser artist it can lead to lots of wrong notes. Rollins's ear in this respect is remarkably dependable.

Dynamically, too, Rollins is a master of contrast and coloring. Listening to "Blue 7" from this point of view is very interesting. There is a natural connection between the character of a given phrase and its dynamic level (in contrast to all too many well-known players who seem not to realize that to play seven or eight choruses resolutely at the same dynamic level is the best way to put an audience to sleep). Rollins's consummate instrumental control allows him a range of dynamics from the explosive outbursts with which he slashes about, for instance, after Max's solo (or later when he initiates the "fours") to the low B natural three bars from the end, a low note which Sonny floats out with a breathy, smoky tone that should make the average saxophonist envious. Rollins can honk, blurt, cajole, scoop, shrill—whatever the phrase demands without succumbing to the vulgar or obnoxious. And this is due largely to the fact that Sonny Rollins is one of those rather rare individuals who has both taste and a sense of humor, the latter with a slight turn toward the sardonic.

Rhythmically, Rollins is as imaginative and strong as in his melodic concepts. And why not? The two are really inseparable, or at least should be. In his recordings as well as during several evenings at Birdland recently Rollins indicated that he can probably take any rhythmic formation and make it swing. This ability enables him to run the gamut of extremes—from almost a whole chorus of nonsyncopated quarter notes (which in other hands might be just naïve and square but through Rollins's sense of humor and superb timing are transformed into a swinging line) to asymmetrical groupings of fives and sevens or between-the-beat rhythms that defy notation.

As for his imagination, it is (as already indicated) prodigiously fertile. It can evidently cope with all manner of material, ranging from Kurt Weill's "Moritat" and the cowboy material of his *Way Out West* Lp (Contemporary 3530) to the more familiar area of ballads and blues. But to date his most successful and structurally unified efforts have been based on the blues. ("Sumphin'," for instance, made with Dizzy Gillespie (Verve 8260) is almost on the level of "Blue 7"; it falls short, comparatively, only in terms of originality, but is also notable for a beautifully organized Gillespie solo.) This is not to say that Rollins is incapable of achieving thematic variations in non-blues material. Pieces such as "St. Thomas" or "Way Out West" indicate more than a casual concern with this problem; and in a recent in-the-flesh rendition of "Yesterdays," a lengthy solo cadenza dealt almost exclusively with the melodic line of this tune. His vivid imagination not only permits him the luxury of seemingly endless variants and permutations of a given motive, but even enables him to emulate ideas not indigenous to his instrument, as for instance in "Way Out West" when Rollins, returning for his second solo, imitates Shelly Manne's closing snare drum roll on the saxophone!

Lest I seem to be overstating the case for Rollins, let me add that both his live and recorded performances do include average and less coherent achievements—even an occasional wrong note, as in "You Don't Know What Love Is" (Prestige LP 7079)—which only proves that (fortunately) Rollins is human and fallible. Such minor blemishes are dwarfed into insignificance by the enormity of his talent and the positive values of his great performances. In these and especially "Blue 7," what Sonny Rollins has added conclusively to the scope of jazz improvisation is the idea of developing and varying a *main* theme, and not just a secondary motive or phrase which the player happens to hit upon in the course of his improvisation and which in itself is unrelated to the "head" of the composition. This is not to say that a thematically related improvisation is *necessarily* better than a free harmonically based one. Obviously any generalization to this effect would be unsound: only the quality of a specific musician in a specific performance can be the ultimate basis for judgment. The point is not—as some may think I am implying—that, since Rollins does a true thematic variation, he therefore is superior to Parker or Young in a nonthematic improvisation. I am emphasizing primarily a *difference* of approach, even though, speaking quite subjectively, I may feel the Rollins position to be ultimately the more important one. Certainly it is an approach that inherently has an important future.

The history of classical music provides us with a telling historical precedent for such a prognosis: after largely non-thematic beginnings (in the early Middle Ages), music over a period of centuries developed to a stage where (with the great classical masters) thematic relationships, either in a sonata or various variational forms, became the prime building element of music, later to be carried even further to the level of continuous and complete variation as implied by Schoenberg's twelve-tone technique: in short, an over-all lineage from free almost anarchical beginnings to a relatively confined and therefore more challenging state. The history of jazz gives every indication of following a parallel course, although in an extraordinarily condensed form. In any case, the essential point is not that, with thematically related solos, jazz improvisation can now discard the great tradition established by the Youngs and Parkers, but rather that by building *on* this tradition and enriching it with the new element of thematic relationships, jazz is simply adding a new dimension. And I think we might all agree that renewal through tradition is the best assurance of a flourishing musical future.

14

Lee Konitz

A "liner note" for a remarkable album recorded by the great alto saxophonist in 1967 on the Milestone label. (MSP 9013)

LEE KONITZ is many things. He is a veteran of the original bop revolution of the early and mid-1940s. He is that rarity, a stylistic innovator whose original contribution some twenty years ago consisted of fusing the musical conceptions of Charlie Parker and Lennie Tristano, and in the process becoming the most successful translator of Tristano's piano-based polyphonic style to the saxophone. He is that other rarity, a totally dedicated musician, uncompromising and tenacious in the face of adversity, a man who has not lost faith in the original idealism which generated the excitement and ferment of the early bop days, and which first drew him to jazz. Among his many achievements one would have to list his work within the Gil Evans arrangements for the great 1948 Claude Thornhill band, his by now "classic" solos on the Miles Davis Nonet (*Birth of the Cool*) recordings, and his collaboration with Bill Russo a decade ago in an album entitled *Image*. He has survived many difficult years, keeping the strain of his thinking pure and clear; and he has with this album made a recording which brilliantly summarizes his talents, his contributions of the past, and his growth through the years.

The format of this recording, conceived by Lee and in itself an ingenious idea, effectively stimulates his musical imagination and resources. By pairing himself off with a number of different musicians, all masters in their own right, Lee automatically solves the whole problem of how to create timbral, textural, and structural variety in some forty-five minutes of music. Lee is thus neither faced with carrying the full burden of an entire LP, nor of having to find within a single instrumental context the variety and contrast so necessary to the enjoyment of large musical statements—a problem the LP created but which unhappily very few musicians have been able to overcome successfully.

The level of playing is remarkably consistent, and Lee and his partners avoid the pitfall of most jazz duet-improvisations: the tedium of contrapuntal and canonic imitations. Whether Lee is providing the musical stimulus or whether he is reacting to one of his colleagues, the inspirational level remains high throughout. This is due at least in part to Lee's insistence that the entire album be recorded in one five-hour session to ensure maximum spontaneity and intensity.

The variety inherent in the format of changing duet-partners is underscored further by the selection of compositions, contrasting not only in tempo and mood, but in style ranging from the very tonal "Tickle Toe" to completely atonal "free" improvisations like "Erb" and the quartet section of "Alone Together."

The highlights of the album are many and varied. "Struttin' with Some Barbecue," a Louis Armstrong vehicle of the 1920s, is imbued with new life by the smooth, almost casual conversational interchange between Lee and Marshall Brown's valve trombone. The track ends with one of several examples of creatively used electronic overdubbing: Lee and Marshall play Armstrong's famous solo on "Struttin' " in octaves, superimposed on a separately recorded stop-time background, played on baritone sax and baritone horn.

On "You Don't Know What Love Is" we find Lee in a complex dialogue with Joe Henderson, a highly regarded new talent on tenor saxophone. Here the lines become more enmeshed and intertwined than on the previous track, a feature which is brought into even sharper focus by the fact that Lee's alto tone here is very rich and dark, and nearly matches the brightish tenor tone of Henderson. Thus the exchange occurs not only on the level of musical ideas but in the subtler realm of sonority (tone-color) as well.

Side One ends with a series of "variations" on "Alone Together." A short introductory statement by Lee on electric alto sax leads to an extraordinary exchange between Elvin Jones and Lee, where Elvin's relentlessly fertile imagination supplies the main thrust. Sounding like a whole African drum ensemble all by himself, Jones is precise, relaxed, sharp, and controlled all at once. The German vibraharp virtuoso, Karl Berger, is Lee's next partner. Berger is the first vibraphonist to really break away from the Milt Jackson sound and style, and his bright, clear, almost glass-like sonority matches Lee's light alto perfectly. This track also features the use of percussive clicking sounds, produced by very close miking of the keys on Lee's alto being manipulated without blowing air into the instrument. Eddie Gomez, a young bassist in the Scott LaFaro tradition, collaborates next in a fine duet, including several bowed passages, in which Gomez's fine tone and refined bowing technique can be admired.

The "Alone Tegother" variations conclude with a quartet improvisation which is not only a highpoint of the record but surely must count as one of the finest "free" improvisations yet recorded. A solo introduction by Elvin Jones sets the pace and leads to the full quartet with Lee intoning the melody of Schwartz's lovely standard as a sort of *cantus firmus,* while Berger and Gomez provide *ostinato*-type embellishments around Lee's "theme," and Jones providing a clear, concise and complex rhythmic substructure. Berger here is especially inventive, using among other things a *staccato* clipped sound on the vibes, produced by stopping the sound with the mallets immediately after the metal plates have been struck. This is a relatively new technique used by various vibraphonists in both the contemporary "classical" and jazz fields, but here exploited with great sensitivity and imagination. The piece progresses as Lee begins to break the theme into increasingly shorter motivic fragments and repeated riff-like figures, the players' responses to each other thus forming into a pattern of quicker reactions and shorter phrases. Repeated listenings to this variation will show to what degree cohesiveness and formal control can be achieved in "free" non-tonal improvisations.

"Checkerboard" features Dick Katz and Lee in Katz's own composition. The two players obviously inspire each other in the manner and tradition (though not the style, of course) of Earl Hines and Louis Armstrong four decades ago on "Weather Bird." Dick Katz emerges in one of his best recorded efforts and here reveals his admiration for Art Tatum and Duke Ellington along with touches—unconscious perhaps—of Lennie Tristano. These sources, however, are welded into a single style which is dominated by Katz's own lean, linear concept of the piano.

"Erb" is named after Jim Hall's teacher of many years ago at the Cleveland Institute, the highly regarded composer Donald Erb. It is an impressionistic piece, moving from a quiet delicate opening to a more agitated middle section, through which Jim Hall leads Lee in a manner that can only be described as spectacular and subtle at the same time. Again the pitchless clicking of the keys of Lee's horn is used. "Erb" demonstrates how organic growth can be achieved in a free improvisation, for Jim brought to the studio no thematic material but merely a graph containing instructions as to dynamics and range.

"Tickle Toe," with Richie Kamuca as Lee's partner, finds both players on tenor sax. Sticking close to Lester Young's theme and his own famous rendition of it, the two men sound so much alike that, unless forewarned, one would assume the recording to be an overdubbed duet between Lee and himself. The two players make it an official tribute to Prez by quoting the latter's familiar solo in unison.

"Duplexity" is a long conversation between Lee on tenor and the Duke Ellington veteran Ray Nance on violin.

The final track is based on an outline provided by Marshall Brown, using all the players (except Nance) in full ensemble, plus occasional overdubbing by Lee on his "varitone" electric sax. The result at the climax is a dense Ivesian multi-layered polyphony. Elvin Jones's entrance (after a brief vamp-interlude for the guitar) must be heard to be believed. It is one of those tiny upbeat explosions in which Elvin excels, which packs enough power to fuel the whole ensemble for several choruses. The side and the album end with Lee trailing off into silence on his "varitone" sax.

On this record Lee Konitz dares much and achieves much. Given the format of ten unaccompanied duets and two ensemble pieces, Lee ventures to expose his musical soul in a kind of nakedness (i.e., without benefit of rhythm sections, manufactured ensembles, familiar tunes, and other safeguards) that few would dare. He takes on his various partners with assurance and equanimity, giving and taking like a master diplomat, in the process proving that *any* combination of instruments can combine effectively when the musicians are as sensitive and imaginative as those assembled here. To these various challenges Lee Konitz rises easily and successfully. And he probably had a ball doing it!

15

The Divine Sarah

A deeply felt tribute to the great singer Sarah Vaughan, presented in the Hall of Flags at the Smithsonian Museum of American History on November 5, 1980. Schuller's paean was followed by a brief concert by Ms. Vaughan in which, according to witnesses, she demonstrated and confirmed every point he had made—all the more remarkable since she did not actually hear any of Schuller's comments.

WHAT I AM ABOUT to do really can't be done at all, and that is to do justice to Sarah Vaughan in words. Her art is so remarkable, so unique that it, *sui generis,* is self-fulfilling and speaks best on its own musical artistic terms. It is—like the work of no other singer—self-justifying and needs neither my nor anyone else's defense or approval.

To say what I am about to say in her very presence seems to me even more preposterous, and I will certainly have to watch my superlatives, as it will be an enormous temptation to trot them all out tonight. And yet, despite these disclaimers, I nonetheless plunge ahead toward this awesome task, like a moth drawn to the flame, because I want to participate in this particular long overdue celebration of a great American singer and share with you, if my meager verbal abilities do not fail me, the admiration I have for this remarkable artist and the wonders and mysteries of her music.

No rational person will often find him or herself in a situation of being able to say that something or somebody is *the best*. One quickly learns in life that in a richly competitive world—particularly one as subject to subjective evaluation as the world of the arts—it is dangerous, even stupid, to say that something is without equal and, of course, having said it, one is almost always immediately challenged. *Any* evaluation—except perhaps in certain sciences where facts are truly incontrovertible—any evaluation is bound to be relative rather than absolute, is bound to be conditioned by taste, by social and educational backgrounds, by a host of formative and conditioning factors. And yet, although I know all that, I still am tempted to say and will now dare to say that Sarah Vaughan is quite simply the greatest vocal artist of our century.

Perhaps I should qualify that by saying the most *creative* vocal artist of our time. I think that will get us much closer to the heart of the matter, for Sarah Vaughan is above all that rare rarity: a jazz singer. And by that I mean to emphasize that she does not merely render a song beautifully, as it may have been composed and notated by someone else—essentially a *re*-creative act—but rather that Sarah Vaughan is a composing singer, a singing composer, if you will, an improvising singer, one who never—at least in the last 25 years or so—has sung a song the same way twice: as I said a *creative* singer, a jazz singer.

And by using the term jazz I don't wish to get us entrapped in some narrow definition of a certain kind of music and a term which many musicians, from Duke Ellington on down, have considered confining, and even denigrating. I use the word "jazz" as a handy and still widely used convenient descriptive label; but clearly Sarah Vaughan's singing and her mastery go way beyond the confines of jazz.

And if I emphasize the creativity, the composer aspect of her singing, it is to single out that rare ability, given, sadly, to so few singers, including, of course, all those in the field of classical music. It is my way of answering the shocked response among some of you a few moments ago when I called Sarah Vaughan the greatest singer of our time. For it is one thing to have a beautiful voice; it is another thing to be a great musician—often, alas, a *truly* remote thing amongst classical singers; it is still another thing, however, to be a great musician with a beautiful and technically perfect voice, who also can compose and create extemporaneously.

We say of a true jazz singer that they improvise. But let me assure you that Sarah Vaughan's improvisations are not mere embellishments or ornaments or tinkering with the tune; they are compositions in their own right or at least re-compositions of someone else's material—in the same manner and at the same level that Louis Armstrong and Charlie Parker and other great jazz masters have been creative.

You can imagine that I do not say these things lightly, and that I do not make so bold as to make these claims without some prior thought and reason. For I am, as many of you know, someone who played for fifteen years in the orchestra of the Metropolitan Opera, loved every minute of it, and during those years heard a goodly share of great singing—from Melchior to Björling and DiStefano, from Flagstad to Sayao to Albanese and Callas, from Pinza to Siepi and Warren. Before that, as a youngster, I thrilled to the recordings of Caruso, Rethberg, Ponselle, Muzio, Easton, and Lawrence. So I think I know a little about that side of the singing art. And yet with all my profound love for those artists and the great music they made, I have never found anyone with the kind of total command of all aspects of their craft and art that Sarah Vaughan has.

I do not wish to engage in polemical discussion here. Nor am I Sarah Vaughan's press agent. I would claim, however—along with Barbara Tuchman—that though my judgment may be subjective, the condition I describe is not. What is that condition? Quite simply a perfect instrument attached to a musician of superb musical instincts, capable of communicating profoundly human expressions and expressing them in wholly original terms.

First the voice. When we say in classical music that someone has a "perfect voice" we usually mean that they have been perfectly trained and that they use their voice seemingly effortlessly, that they sing in tune, produce not merely a pure and pleasing quality, but are able to realize through the proper use of their vocal organs the essence and totality of their natural voice. All that can easily be said of Sarah Vaughan, leaving aside for the moment whether she considers herself to have a trained voice or not. As far as I know, she did study piano and organ, but not voice, at least not in the formal sense. And that may have been a good thing. We have a saying in classical music—alas, painfully true—that given the fact that there are tens of thousands of bad voice teachers, the definition of a great singer is one who managed *not* to be ruined by his or her training. It is better, of course, to be spared the taking of those risks.

There is something that Sarah Vaughan does with her voice which is quite rare and virtually unheard of in classical singing. She can color and change her voice at will to produce timbres and sonorities that go beyond anything known in traditional singing and traditional vocal pedagogy. (I will play, in a while, a recorded excerpt that will show these and other qualities and give you the aural experience rather than my—as I said earlier—inadequate verbal description.)

Sarah Vaughan also has an extraordinary range, not I hasten to add used as a gimmick to astound the public (as is the case with so many of those singers you are likely to hear on the Tonight Show), but totally at the service of her imagination and creativity. Sarah's voice cannot only by virtue of its range cover four types of voices—baritone, alto, mezzo soprano, and soprano, but she can color the timbre of her voice to emphasize these qualities. She has in addition a complete command of the effect we call *falsetto,* and indeed can on a single note turn her voice from full quality to *falsetto* (or, as it's also called, head tone) with a degree of control that I only heard one classical singer ever exhibit, and that was the tenor Giuseppe Di-Stefano—but in his case only during a few of his short-lived prime years.

Another thing almost no classical singers can do and something at which Sarah Vaughan excells is the controlled use of vibrato. The best classical singers develop a vibrato, of a certain speed and character, which is nurtured as an essential part of their voice, indeed their trademark with the public, and which they apply to all music whether it's a Mozart or Verdi opera or a Schubert song. Sarah Vaughan, on the other hand, has a complete range, a veritable arsenal of vibratos, ranging from none to a rich throbbing, almost at times excessive one, all varying as to speed of vibrato and size and intensity—at will. (Again my recorded example will demonstrate some truly startling instances of this.)

Mind you, what Sarah Vaughan does with the controlled use of vibrato and timbre was once—a long time ago—the *sine qua non* of the vocal art. In the seventeenth and eighteenth centuries vibrato, for example, was not something automatically used, imposed, as it were, on your voice. On the contrary, it was a special effect, a kind of embellishment—an important one—which you used in varying degrees or did not use, solely for various expressive purposes and to heighten the drama of your vocal expressivity. It is an art, a technique which disappeared in the nineteenth century and is all but a lost art today, certainly amongst classical singers, who look at you in shocked amazement if you dare to suggest that they might vary their vibrato or timbre. They truly believe they have *one* voice, when potentially—they don't realize it—they could (should) have several or many.

Here again, I think Sarah learned her lessons not from a voice teacher, but from the great jazz musicians that preceded her. For among great jazz instrumentalists the vibrato is not something sort of slapped onto the tone to make it sing, but rather a compositional, a structural, an expressive element elevated to a very high place in the hierarchy of musical tools with which they express themselves.

Another remarkable thing about Sarah Vaughan's voice is that it seems ageless; it is to this day perfectly preserved. That, my friends, is a sign—the only sure sign—that she uses her voice absolutely correctly, and will be able to sing for many years more—a characteristic we can find, by the way, among many popular or jazz singers who were *not* formally voice-trained. Think of Helen Humes, Alberta Hunter, Helen Forrest, Chippie Wallace, Tony Bennett, and Joe Williams.

So much for the voice itself. Her musicianship is on a par with her voice and, as I suggested earlier, inseparable from it. That is, of course, the ideal condition for an improvising singer—indeed a prerequisite. For you cannot improvise, compose extemporaneously, if you don't have your instrument under full control; and by the same token, regardless of the beauty of your voice, you have to have creative imagination to be a great jazz or improvising singer. Sarah's creative imagination is exuberant. I have worked with Sarah Vaughan, I have accompanied her, and can vouch for the fact that she never repeats herself or sings a song the same way twice. Whether she is using what we call a paraphrase improvisation—an enhancement of the melody where the melody is still recognizable—or whether she uses the harmonic changes at the basis of the song to improvise totally new melodies or gestures, Sarah Vaughan is always totally inventive. It is a restless compulsion to create, to reshape, to search. For her a song—even a mediocre one—is merely a point of departure from which she proceeds to invent, a skeleton which she proceeds to flesh out.

There are other singers—not many—who also improvise and invent, but I dare say none with the degree of originality that Sarah commands. She will come up with the damndest musical ideas, unexpected and unpredictable leaps, twisting words and melodies into new and startling shapes, finding the unusual pitch or nuance or color to make a phrase uniquely her own. When one accompanies her one has to be solid as a rock, because she is so free in her flights of invention that she could throw you if you don't watch out. She'll shift a beat around on you, teasing and toying with a rhythm like a cat with a mouse, and if you're not secure and wary, she'll pull you right under. She is at her best and her freest when her accompaniment is firmly anchored.

Perhaps Sarah Vaughan's originality of inventiveness is her greatest attribute, certainly the most startling and unpredictable. But unlike certain kinds of unpredictability—which may be merely bizarre—Sarah's seems immediately, even on first hearing, inevitable. No matter how unusual and how far she may stretch the melody and harmony from its original base, in retrospect one senses what she has just done as having a sense of inevitability—"Of course, it had to go that way; why didn't I think of that?" I go further: in respect to her originality of musical invention I would say it is not only superior to that of any other singer, but I cannot think of any active jazz instrumentalist—today—who can match her.

If it is true, as has often been stated through the centuries, that one way of defining high art is by the characteristic of combining the expected with the unexpected, of finding the unpredictable *within* the predictable, then Sarah Vaughan's singing consistently embodies that ideal.

Lastly, I must speak of the quality of Sarah's expressiveness, the humanism, if you will, of her art. Sarah has a couple of nicknames, as some of you know. The earliest one was Sassy. Next, around the early 1950s, she came to be called "the Divine Sarah," and more recently simply "the Divine One." Now that's a lovely thing to say about anyone, and I would not argue about Sarah's musical divinity, except in one somewhat semantic respect. What I love so in her singing is its humanness, its realness of expression, its integrity. It is nice to call her singing divine, but it's more accurate to call it human. Under all the brilliance of technique and invention, there is a human spirit, a touching soul, and a gutsy integrity that moves us as listeners.

How does one measure an artist's success? By how much audience they attract? By how much money they make? By how many records they sell? Or by how deeply they move a sophisticated or cultured audience? Or by how enduringly their art will survive? Sarah has been called the musicians' singer—both a wonderful compliment and a delimiting stigmatization. What seems to be true for the moment is that her art, like Duke Ellington's, is too subtle, too sophisticated to make it in the big—really big—mass pop market. God knows, Sarah—or her managers—have tried to break into that field. But she never can make it or will make it, like some mediocre punk rock star might, because she's too good. She can't resist being inventive; she can't compromise her art; she must search for the new, the untried; she must take the risks.

And she will be—and is already—remembered for *that* for a long time. To some like me—I've been listening to her since she was the very young, new girl singer with the Billy Eckstine Band in the mid-1950s—she is already a legend. I invite you now to listen to the promised excerpt—only *one* example of her art—a stunning example indeed, taken from a 1973 concert in Tokyo, during which Sarah Vaughan sang and recomposed *My Funny Valentine*. Listen!!

(record played)

It is now my privilege to exit gracefully and to invite you to listen to the one and only Sarah Lois Vaughan!

16

Gil Evans

Schuller's Grove *entry on jazz's most outstanding arranger-composer-recomposer of the last four decades. Schuller frequently played horn for Gil Evans in his earlier horn-playing days* (Birth of the Cool, Porgy and Bess) *and found it difficult to couch his admiration for Evans in the formal, "objective" language required by the* New Grove Dictionary of American Music.

EVANS, GIL [Green, Ernest Gilmore] (b Toronto, 13 May 1912). Canadian jazz arranger, composer, pianist, and band-leader. A self-taught musician, he led his own band in Stockton, California, from 1933 to 1938. When the singer Skinnay Ennis took over the band in the latter year, Evans stayed on as arranger. In 1941 he joined Claude Thornhill in the same capacity, contributing in 1947 such outstanding arrangements as "Anthropology," "Donna Lee," "Yardbird Suite," "The Old Castle," "Robbins Nest," and "A Sunday Kind of Love." In these works and others of the period Evans used, in addition to the standard swing era big-band instrumentation, two French horns and a tuba, which, along with the restrained control of vibrato in the saxophones and brass, produced a rich, dark-textured "cool" orchestral sound, anticipated only by Duke Ellington and Eddie Sauter. In their emphasis on ensemble over improvised solo, Evans's scores for Thornhill, rather than mere arrangements, were in essence "recompositions" and "orchestral improvisations" on the original materials, e.g. Charlie Parker lines, popular songs, classical works (such as Moussorgsky's *Pictures at an Exhibition*).

From 1948 to 1950 Evans contributed prominently to the Miles Davis Nonet recordings (for Capitol, known later as *Birth of the Cool*). In his memorable scores "Boplicity" and "Moon Dreams," Evans captured the essential sound and texture of the Thornhill band in a reduced-ensemble format. Oddly, his work for both Davis and Thornhill was roundly ignored by critics and jazz audiences alike.

After a period of relative obscurity, during which Evans worked in radio and television, he returned to jazz with three memorable LP recordings, all written for and featuring Miles Davis: *Miles Ahead* (1957), *Porgy and Bess* (1959), and *Sketches of Spain* (1960). In these, as well as *New Bottle, Old Wine* (1958), Evans extended his earlier orchestral concepts to larger instrumental forces (up to 20), often achieving a distinctive synthesis of varied timbral mixtures, opaque almost cluster-like voicings alternating with rich polyphonic textures, all couched in an advanced harmonic language.

Since the early 1960s Evans has made several attempts to form permanent orchestras, but these have been unable to establish themselves, resulting occasionally, however, in excellent recordings: *The Individualism of Gil Evans* (1963-64), *Blues in Orbit* (1969-71), *Priestess* (1977). In recent years Evans has incorporated electrified instruments (piano, bass, synthesizer, etc.) into his ensembles, along with a noticeable tendency to integrate more extended solo spaces into his arrangements and compositions. This has led to a considerable loosening up of Evans's style, both in respect to form and texture, compared with the more compact and veiled densities of his earlier arrangements. Even so the overall temper of his work remains moody, poignant, and introverted, reflected in a distinct predilection for pieces set in minor keys.

Although initially influenced by the middle-period works of Duke Ellington, Evans developed a style wholly his own, memorable especially for its richly chromatic though always tonally-oriented harmonic language and its seemingly inexhaustible blendings of instrumental timbres. No mere coloristic effects, they are very often the very substance of his art, providing in turn imaginative frameworks for his soloists in ways equaled in the history of jazz only by Morton, Ellington, and Mingus. Evans has succeeded in preserving, even in his most elaborate scores, the essential spontaneity and improvisatory nature of jazz, achieving a rare symbiotic relationship between composed and improvised elements.

Although known chiefly as an arranger, Evans has in recent years increasingly devoted himself to composition. His more notable works include "Flute Song," "Las Vegas Tango," "Proclamation," "Variations on the Misery," "Anita's Dance," and (co-composed with Miles Davis) "Hotel Me" and "General Assembly."

17

Alec Wilder

Schuller's entry on America's remarkable maverick composer for the New Grove Dictionary of American Music. *Schuller now publishes most of Wilder's instrumental works in his publishing company, Margun Music, Inc.*

WILDER, ALEC [Alexander] (Lafayette Chew) (b. Rochester, New York, 16 Feb. 1907; d. 24 Dec. 1980). American composer, arranger. After private (non-degree) studies at the Eastman School, he became active in the early 1930s as a songwriter and arranger in New York. Many of his songs were composed for and/or performed by Mildred Bailey, Cab Calloway, Bing Crosby, Ethel Waters, Mabel Mercer, and (in the 1940s) Frank Sinatra. In 1939 Wilder attracted attention with a series of octets (with whimsical titles such as "Sea Fugue Mama," "Neurotic Goldfish," "Poltergeist," "The Home Detective Registers"). Scored for winds and rhythm section (including harpsichord) these airy, elusive pieces blended popular melodies and swing rhythms with classically oriented forms.

In the early 1950s, Wilder turned from the world of popular songs to writing chamber and orchestral music, eventually also operas. These 300-odd compositions, written for almost every conceivable instrumental combination, are characterized by his unique melodic gift, a harmonic language alternating between French impressionism and modal (often fugal) writing, and a preference for loosely linked Suite forms.

Although much admired and performed by certain musicians on both sides of the musical fence (jazz: Stan Getz, Gerry Mulligan, Marian McPartland, Roland Hanna; classical: John Barrows, Bernard Garfield, Harvey Phillips, Gary Karr), Wilder's "eclectic" style was largely rejected by both musical establishments.

An unclassifiable "American original," he drew upon a wide variety of personal musical influences. In his best works, he was able to forge a style uniquely his own, characterized by those elements he most cherished in other composers: unclutteredness, honest sentiment, unexpectedness, singing melodies, and sinuous phrases.

Wilder also wrote (in collaboration with James T. Maher) *American Popular Song* (Oxford University Press, 1972), a lovingly insightful study of the subject covering the period between 1900 and 1950, and *Letters I Never Mailed* (Little, Brown, 1975).

It is perhaps at this time still too early to render a full assessment of Wilder's *oeuvre,* much of it remaining un- or rarely performed and only recently beginning to be available in published form.

OPERA: *Miss Chicken Little,* A Musical Fable (1954); *Ellen* (1955); *The Opening* (1972); *The Truth About Windmills* (1975); *Kittiwake Island* (1955); *The Lowland Sea* (1952); *Sunday Excursion* (1953); *Cumberland Fair* (1953).

MUSICAL COMEDY: *Jack in the Country.*

BALLET: *Juke Box* (1940); *Three Ballets in Search of a Dancer; False Dawn; The Green Couch.*

ORCHESTRA/WIND ENSEMBLE: Numerous Concertos, Suites, Entertainments, Airs; *Serenade for Winds* (1977), *A Child's Introduction to the Orchestra* (1954); *Carl Sandburg Suite* (1960); *Grandma Moses Suite* (1950).

CHAMBER MUSIC: Several hundred works for woodwinds, brass, and strings, ranging from solos, duos, trios, quartets, quintets, and octets, in many diverse combinations (including standard formats—such as 12 Woodwind Quintets, 8 Brass Quintets).

PIANO: *Pieces for Young Pianists,* Vol. 1 and 2; *Six Suites; Un Deuxième Essai* (1965).

SONGS: Numerous popular songs, most notably "While We're Young" (1943): "I'll Be Around" (1942); "It's So Peaceful in the Country" (1941); "The Winter of My Discontent" (1955); "Baggage Room

Blues" (1954); "Blackberry Winter" (1976); also several dozen art songs.

PRINCIPAL PUBLISHERS: Margun Music (instrumental music, operas); Ludlow Music/The Richmond Organization (popular songs); Associated Music Publishers; Kendor.

BIBLIOGRAPHY: Current Biography, Vol. 41, July 1980. Whitney Balliett, *Alec Wilder and His Friends,* Houghton Mifflin (1974).

18

Third Stream

The first written formal statement on Schuller's Third Stream concept, originally limited to a fusion of jazz and contemporary "classical" music techniques (in the intervening years expanded to embrace all manner of folk, ethnic, vernacular, non-Western musics as well). The article was published in the Saturday Review of Literature *(May 13, 1961) when the partisan arguments pro and con Third Stream raged hot and furious, especially on the part of the jazz critical fraternity, many of whom felt threatened by this "hybrid intruder."*

ABOUT A YEAR AGO the term "Third Stream," describing a new genre of music located about halfway between jazz and classical music, attained official sanction by use in a headline of the *New York Times.* Since that day, almost everybody has had his say about this music and its practitioners, and, not surprisingly, much nonsense about it has found its way into print. I suppose this is inevitable. If we consider how much confusion and prejudice are attached to jazz and contemporary music as *separate* fields of musical activity, we can then imagine how existing misunderstandings are likely to be compounded when we speak of a *fusion* of these two forces.

I first used the term "Third Stream" in a lecture three or four years ago, in an attempt to describe a music that was beginning to evolve with growing consistency. For lack of a precise name, one was forced at the time to describe it either in a lengthy definition or in descriptive phrases comprising several sentences. I used the term as an adjective, not as a noun. I did not envision its use as a name, a slogan, or a catchword (one thinks back with horror to the indignities visited upon the modern jazz movement in the late forties by the banal commercialization of the catchword "Bop"); nor did this imply a sort of "canonization," as one critic facetiously put it; nor, least of all, did I intend the term as a commercial gimmick. Such a thought evidently comes most readily to some people's minds in a society in which commercial gimmickry is an accepted way of life. Ultimately, I don't care whether the term "Third Stream" survives. In the interim it is no more than a handy descriptive term. It should be obivous that a piece of Third Stream music is first of all *music,* and its quality cannot be determined solely by categorization. Basically I don't care what category music belongs to; I only care whether it is good or bad. As one fellow musician put it: "I like jazz, not because it's jazz, but because it's good music."

After a year of watching the confusion mount and the increasing commercial exploitation of the phrase "Third Stream" (often incorrectly used, to boot), and after reading numerous reviews revealing the writers' calloused misunderstanding of the music involved, I somehow feel obligated—since I may have indirectly instigated all this nonsense—finally to have *my* say about it, and in the process I hope to bring some clarity to the whole issue.

I am fully aware that, individually, jazz and classical music have long, separate traditions that many people want to keep separate and sacred. I also recognize the right of musicians in either field to focus their attention entirely on preserving the idiomatic purity of these traditions. It is precisely for these reasons that I thought it best to separate from these two traditions the new genre that attempts to fuse "the improvisational spontaneity and rhythmic vitality of jazz with the compositional procedures and techniques acquired in Western music during 700 years of musical development." I felt that by designating this music as a *separate, third* stream, the two other mainstreams could go their way unaffected by attempts at fusion. I had hoped that in this way the old prejudices, old worries about the purity of the two main streams that have greeted attempts to bring jazz and "classical" music together could, for once, be avoided. This, however, has not been the case. Musicians and critics in both fields have considered this Third Stream a frontal attack on their own traditions.

Characteristically, the jazz side has protested against the intruder more vigorously than its opposite partner. And since my music in this genre has by now been accused of everything from "opportunism" to "racial callousness," I had better make some points unequivocally clear. I can best clarify my intentions by some straightforward categorical statements: 1) I am not interested in *improving* jazz (jazz is a healthy young music, and I would not presume to be capable of improving it); 2) I am not interested in *replacing* jazz (a thought like this could only emanate from some confused believer in the basic inferiority of jazz); 3) I am not interested in "bringing jazz into classical music," nor in "making a lady out of jazz," in the immortal words of the erstwhile "symphonic jazz" proponents (Paul Whiteman, Ferde Grofé, et al., come to mind).

I am simply exercising my prerogative as a creative artist to draw upon those experiences in my life as a musician that have a vital meaning for me. It is inevitable that the creative individual will in some way reflect in his creative activity that which he loves, respects, and understands; my concern, therefore, is precisely to preserve as much of the essence of both elements as is possible.

Yet, with disturbing consistency, "jazz critics" keep appraising the Third Streamers solely on the basis of jazz. The producers of this music are also equally at fault. One recently issued recording describes itself as "Third Stream Jazz," and I suppose before long we will have some "Third Stream Classics"—whatever that might turn out to be. This is analogous to calling a nectarine a nectarine-plum. The standards of jazz, per se, are not applicable to Third Stream, any more than one can expect a nectarine to taste like a peach. A Third Stream work does not wish to be heard as jazz alone; it does not *necessarily* expect to "swing like Basie" (few can, even within the jazz field); it does not expect to seduce the listener with ready-made blue-noted formulas of "soul" and "funk"; and it certainly does not expect to generate easy acceptance among those whose musical criteria are determined only on the basis of whether one can snap one's fingers to the music. And, I might add, it is not an attempt to find a niche in the hearts of the American public by capitalizing on the brash, racy "jazziness" of some recent Broadway musicals.

In *my* understanding of the term, Third Stream music must be born out of respect for and full dedication to *both* the musics it attempts to fuse. (This is more than one can say for the pop song or rock-'n'-roll commercializers of jazz, about whom, ironically, I have heard no serious complaints.) The lifting of external elements from one area into the other is happily a matter of the past. At its best Third Stream can be an extremely subtle music, defying the kind of easy categorization most people seem to need before they can make up their minds whether they should like something or not. As John Lewis once put it to me in a conversation: "It isn't so much what we see (and hear) in the music of each idiom; it is more what we do *not* see in the one that already exists in the other."

Certainly both musics can benefit from this kind of cross-fertilization, in the hands of gifted people. For example, the state of performance in classical music at the professional level today is, despite all we hear about our skilled instrumentalists, rather low. Most performances touch only the surface of a work, not its essence; and this is most true of the performance of contemporary music. If virtuosic perfection at least were achieved, one could—in a forgetful moment—be satisfied with that. But one cannot even claim this, since the leisurely attitude of the majority of classical players toward rhythmic accuracy is simply appalling, and would seem so to more people were it not so widespread as to be generally accepted. There is no question in my mind that the classical world can learn much about timing, rhythmic accuracy, and subtlety from jazz musicians, as jazz musicians can in dynamics, structure, and contrast from the classical musicians.

It is this kind of mutual fructifying that will, I believe, be one of the benefits of Third Stream. It seems, therefore, unfair for critics to blame a piece of Third Stream music because "the symphony orchestras can't swing." Of course they can't swing; even in their own music they barely manage. But perhaps they will learn; and, when they do, will the compositions involved then be intrinsically better? It is this kind of confusion of levels that I am deploring. It happens, needless to say, on both sides of the fence.

It is not a lame apology but rather a statement of fact to say that this movement is still in its beginnings. The performance problems are still enormous, and much musical adjustment will have to be made by both sides before the compositional ideals of the composers can be realized on the performance level. However, if a symphony orchestra can be made to swing just a little, and if a compositional structure that makes jazz musicians push beyond the thirty-two-bar song forms of conventional jazz can be achieved, are not these already important achievements in breaking the stalemate artificially enforced by people who wish tenaciously to keep the two idioms separate? (That there are not only musical problems involved here, but also deep-rooted racial issues, need hardly be emphasized.)

If historical confirmation is needed, it is easy to point to many precedents, which undoubtedly were met with as much scoffing as Third Stream has encountered. The next time someone invokes the old saw about oil and water not mixing, just tell him that the most sacred and rigorously organized music of all time, the Flemish contrapuntal masses, was more often than not based on what once had been semi-improvised lighthearted, often risqué secular troubadour ballads. Or tell him that the minuet, at first a simple and often crude popular dance, became a sophisticated classical form in the hands of Haydn and Mozart. Or if examples be needed from our own time, Bartók's music—originally compounded of Debussyan and Straussian elements—rose to its greatest heights *after* he was able to fuse his early style with the idiomatic inflections of Eastern European folk music. Many more examples exist.

It seems to me that the kind of fusion which Third Stream attempts is not only interesting but inevitable. We are undergoing a tremendous process of musical synthesis, in which the many radical innovations of the earliest decades of our century are being finally assimilated. Whenever I read one of the "it-can't-be-done" reviews, I think of Wagner's line in *Die Meistersinger,* addressing Beckmesser: "Wollt ihr nach Regeln messen, was nicht nach eurer Regeln Lauf, sucht davon erst die Regeln auf! Eu'r Urtheil, dünkt mich, wäre reifer, hörtet ihr besser zu." Freely translated: "Would you judge by conventional rules that which does not follow those rules? Your judgment, it seems to me, would be more mature if you listened more carefully."

It would seem that the Beckmessers of today are equally incapable of listening to music in terms of a *total* musical experience. When confronted with passages thoroughly fusing the worlds of jazz and classical music, they insist on hearing them in their separate categories, most likely because they can hear well only in one or the other. I am not by nature a polemicist or a crusader, and I would not, in any case, crusade for Third Stream music. I simply would hope for critics (amateur and professional) who could appraise the music on *its* terms, not *theirs.*

* *Third Stream on Discs:* "Modern Jazz Quartet with Orchestra"—Atlantic S 1359; Modern Jazz Quartet: "Third Stream Music"—Atlantic S 1345; Don Ellis: "How Time Passes"—Candid 8004; William Russo: "Image of Man"—Verve; Jimmy Giuffre: "Seven Pieces"—Verve 68307; John Lewis: "Golden Striker"—Atlantic S 1334.

19

Third Stream Revisited

The contents of a September 1981 brochure intended for recruiting purposes by the New England Conservatory, the only school in the United States that maintains a Third Stream department. The pamphlet was to include a visual depiction of the Conservatory as Noah's Ark, taking into its "bosom" all the world's musics—again Schuller's global view of music, representing the ultimate brother/sisterhood of all God's musical creatures.

THIRD STREAM is a way of composing, improvising, and performing that brings musics together rather than segregating them. It is a way of making music which holds that *all musics are created equal,* coexisting in a beautiful brotherhood/sisterhood of musics that complement and fructify each other. It is a global concept which allows the world's musics—written, improvised, handed-down, traditional, experimental—to come together, to learn from one another, to reflect human diversity and pluralism. It is the music of rapprochement, of *entente*—not of competition and confrontation. And it is the logical outcome of the American melting pot: *E pluribus unum.*

As originally articulated by Gunther Schuller, twenty-five years ago, Third Stream was a musical offspring born of the wedlock of two other primary "streams": classical and jazz. It was intended as a handy descriptive *adjective* for a music that already existed but which had no name. It was not intended as a *noun* and certainly not as a slogan, although that's how it came to be viewed once it made headlines in the *New York Times.*

That's when Third Stream's troubles began. How dare this mongrel upstart intrude on our sacred territories! How dare this half-breed contaminate our pure streams!

Admittedly Third Stream is anti-pedigree and anti-establishment—especially when establishment equates with entrenched, inflexible, self-preserving positions and closed minds. Attacks on its name by detractors notwithstanding, Third Stream is also the ultimate anti-label music, for its very essence is based on the concept of diversity and non-categorization. Third Stream is subject to both the benefits and the risks of cross-fertilization. It doesn't always work, but when it does it is something new, rare and beautiful.

And Third Stream is nothing if it is not new and creative; Third Stream is nothing if it dabbles and tinkers; Third Stream is nothing if it fails to amalgamate at the most authentic and fundamental levels. It is not intended to be a music of paste-overs and add-ons; it is not intended to be a music which superficially mixes a bit of this with a bit of that. When it does, it is *not* Third Stream; it is some other nameless kind of poor music.

Third Stream has not only survived; it has flourished and expanded its horizons. From its original idea to fuse classical and jazz concepts and techniques, it has broadened out—in ways that are an apt corollary to our expanded knowledge of non-Western cultures and the rapid shrinking of our globe—to embrace, at least potentially, all the world's ethnic, vernacular, and folk music. It is a non-traditional music which exemplifies cultural pluralism and personal freedom. It is for those who have something to say creatively/musically but who do not necessarily fit the predetermined molds into which our culture always wishes to press us. Third Stream, more than any other concept of music, allows those individuals who, by accident of birth or station, reflect a diversified cultural background, to express themselves in uniquely personal ways. It is in effect *the quintessential American music.*

Oh, yes, we almost forgot.

What Third Stream is not:

1. It is not jazz with strings.
2. It is not jazz played on "classical" instruments.
3. It is not classical music played by jazz players.
4. It is not inserting a bit of Ravel or Schoenberg between be-bop changes—nor the reverse.
5. It is not jazz in fugal form.
6. It is not a fugue performed by jazz players.
7. It is not designed to do away with jazz or classical music; it is just another option amongst many for today's creative musicians.

And by definition there is no such thing as "Third Stream jazz."

See you at the Conservatory.

20

The Avant-Garde and Third Stream

A brief history and introduction to the subject, commissioned for New World Records album Mirage *(NW 216).*

THERE IS A VIEW, shared by many jazz historians and writers, that the history of jazz parallels in its broad outlines that of Western classical music—only on a much briefer time scale: what took nearly nine centuries in European music is concentrated into a mere six decades in jazz. According to this view, the separate lines of classical music and jazz, veering steadily toward each other from divergent starting points, eventually converge and become one. While this final point has perhaps not yet been reached, both the rapprochement between classical music and jazz and the steady catching up of jazz techniques and concepts with those of the Western avant-garde has brought the two idioms so close that at times they are barely distinguishable from each other.

One could also describe this process of acculturation as the Europeanization of jazz. For while the origins of jazz are certainly traceable to African antecedents, there can be little doubt—although the jazz fraternity doesn't like to admit it—that jazz has already assimilated and transformed countless European musical elements in its brief history. Indeed, this is a process which began with the very beginnings of jazz and ragtime, when these styles were, in themselves, a simple amalgam of African elements with American elements imported from Europe. It is a process which has never ceased, and which reached its apex in the postwar period—extending into the early sixties—in a two-pronged stretching of the confines of jazz: one direction can properly be called the early avant-garde of the late forties and fifties, while the other is the Third Stream movement which began in the late fifties. These two areas are the subject and content of this record.

While many jazz historians have attempted to characterize the development of jazz in purely racial terms, with white musicians and white musical influences always playing the role of the (alleged) corrupters of the "pure" black musical strains, the facts belie such simplistic notions. True, the major innovators of jazz have certainly always been black, and the commercial initiatives and exploitation of jazz have usually come from the white side. Still, it cannot be said that all white or European influences on jazz have been *a priori* negative and corruptive, unless one simply wants to maintain absolute racial purity and hold that *any* multi-ethnic, multi-stylistic fusion is in itself nonproductive.

In any case, whatever anyone may theorize or wish either to prevent or generate tends to be academic, since the course of the music is not normally determined in the academies or by establishment institutions. Rather, the music develops at a grassroots level, is subject to all manner of subtle sociological, economic, and even political pressures, and is often influenced by fads and fashions, by accidents of timing and fate, and by population shifts and other socio-economic factors. In other words, these cross-fertilizations do occur in free and unpredictable patterns, whether anyone approves of them or not.

Beyond the question of their socio-philosophical validity is the further question of the musical integrity of such cultural cross-influences. Here again, that question is mostly irrelevant—as, ultimately, are all questions of artistic pedigree: mongrels are not inherently any the less successful or attractive than a pure breed. Indeed, had cultural traditions never mixed, the last nine hundred years of Western European musical development could never have occurred, because no significant musical innovation has ever been achieved which did not borrow from geographically or stylistically neighboring cultural traditions. Thus, the secular ballads of the troubadours became an essential structural element of the sacred motets of the fourteenth and fifteenth-century *Ars Nova;* and the folk and dance music of the last five centuries has at various times and in various ways profoundly affected the "art music" of composers from Bach and Mozart to Bartók and Stravinsky.

What such cross-influences do is to expand the potential resources of the music. Thus, in jazz, most black musicians in the twenties, though they may be reluctant to admit it now, were eager to emulate the instrumental sophistication and technical control of the Paul Whiteman orchestra. Nor can there be much doubt that in the twenties Ellington was as much inspired and influenced by the advanced harmonic writing of Whiteman's arrangers Ferde Grofé and Bill Challis as he was by the arranging techniques of Will Vodery or the orchestrations of Ravel and Delius (which Ellington is alleged to have listened to—a point never really proven and a "secret" which the Duke, who could be as enigmatic and elusive as anyone, carried with him to his grave). While on the subject of Whiteman, his use of violins—regarded by many as a nefarious and degenerative classical influence—was not at all a handicap to Grofé and Challis, but was rather an additional distinctive musical resource which they exploited eagerly and ingeniously. No other instrument then available in jazz could equal the sustaining ability of the violin, its ease and brilliance in the upper register, and finally its unique timbre. When Charlie Parker recorded with strings in the early fifties, or when Mingus consistently uses a cello, or when Lawrence Brown of the Ellington orchestra emulated the sonority and elegance of movement of the cello on his trombone, these musicians were in their different ways discovering and creatively using resources not normally found in the jazz tradition.

The invasion of Carnegie Hall in 1914 by James Reese Europe with a super-orchestra of 140 filled with enormous numbers of banjos and mandolins, playing classical overtures as well as "syncopated Negro music", Scott Joplin's opera *Treemonisha;* the "symphonic jazz" movement furthered by men like Dave Peyton, Doc Cook, Carrol Dickerson, Wilbur Sweatman, and, of course, Paul Whiteman and George Gershwin; the highly disciplined and carefully worked out and rehearsed performances of groups like Jelly Roll Morton's Red Hot Peppers or Alphonse Trent's orchestra in the twenties; Benny Goodman's conquest of Carnegie Hall in 1938, twenty-five years after James Europe had played there, and Goodman's collaboration with Josef Szigeti and Bela Bartók in the latter's *Contrasts;* Igor Stravinsky's ragtime pieces around the time of World War I and his *Ebony Concerto* composed for Woody Herman some thirty years later; the various nibblings at classical concepts by the likes of John Kirby, Art Tatum, and Artie Shaw; Eddie Sauter's too-advanced (and therefore rarely or never performed) arrangements for Benny Goodman in the late thirties; the increasing fascination with instruments like the French horn, the oboe, and the bassoon, usually found only on the classical side of the tracks; the greater concern for extended forms not associated with the standard jazz forms of the twelve-bar blues or the thirty-two-bar song structure—all these and many more were stations in an ongoing development, a historical continuum which significantly and often positively affected the course of jazz.

The traffic was not only one way, of course. Classical composers were fascinated by the new rhythms and sonorities of jazz too. From Charles Ives, who was the first lonely voice to recognize the musical validity of ragtime, through European composers like Darius Milhaud, Maurice Ravel, Ernst Křenek, Paul Hindemith, Bohuslav Martinu, and Erwin Schulhoff, to Americans like John Alden Carpenter, Louis Gruenberg, Aaron Copland, and William Grant Still—all were captivated by the new fascinations of jazz, although in only a very few instances did they really understand and appreciate the true improvisatory nature of jazz. (This aspect of the cross-fertilization between jazz and classical music will be dealt with on a future New World disc which will contain works by John Alden Carpenter, Henry F. B. Gilbert, Adolf Weiss, and John Powell.)

THE RECORDINGS

SIDE ONE BAND 1

Summer Sequence (Parts 1, 2, 3)
(RALPH BURNS)

Woody Herman and His Orchestra: Sonny Berman, Cappy Lewis, Conrad Gozzo, Pete Candoli, and Shorty Rogers, trumpets; Ralph Pfeffner, Bill Harris, Ed Kiefer, and Lyman Reid, trombones; Woody Herman, clarinet; Sam Marowitz and John LaPorta, clarinets and alto saxophones; Flip Phillips and Micky Folus, tenor saxophones; Sam Rubinowitch, baritone saxophone; Ralph Burns, piano; Chuck Wayne, guitar; Joe Mondragon, bass; Don Lamond, drums. *Recorded September 19, 1946, in Los Angeles. Originally issued on Columbia 38365, 38366, and 38367 (mx #HCO2044, 2055, and 2066).*

By the mid-1940s, with Ellington's superb experiments with form and structure, with a truly orchestral formulation of the traditional jazz instruments, with his development of jazz composition rather than the more or less skillful arrangement of tunes, and with the harmonic/rhythmic innovations of the early bop movement through

Parker, Gillespie, and Monk—already all accomplished—the stage was set for further exploration of these new musical territories. It is a juncture in jazz history which we can readily see as parallel to those years between 1905 and 1909 when classical composers such as Debussy, Mahler, Stravinsky, Schoenberg, and Ives were putting the final touches to the dissolution of tonality and in turn devising new systems of tonally free music. In jazz, the crossing of that threshold can be heard in such compositions as Ralph Burns's *Summer Sequence* and Ellington's remarkable foray into atonality, "Clothed Woman," as well as Tristano's atonal-contrapuntal studies of 1946, "I Surrender Dear" and "I Can't Get Started." The latter two titles are not represented on this LP, but a similar treatment of Jerome Kern's "Yesterdays" dating from 1949 is.

In 1977, thirty years after its creation, one must marvel at the compositional cohesion, craftsmanship, and emotional strength of Ralph Burns's four-movement *Summer Sequence*. (The fourth movement, "Early Autumn," not included here, was composed separately and revised a few years later into a solo vehicle for the tenor saxophonist Stan Getz; in this independent form it survived its three sister movements, perhaps because it contains virtually none of the "foreign" intrusions with which the other movements abound, thus tacitly acceding to broader popular tastes.)

The first nine measures of *Summer Sequence* present in capsuled form the diverse influences at work in this piece. A quiet and lonely duet for two clarinets, classical in conception (it could have been written by Milhaud, Hindemith, or Berg) is joined in the fourth measure by a rising pizzicato figure in bass and piano. A descending sequential phrase of two measures, with harmon muted trumpets subtly introducing the first sign of a true jazz timbre, erupts unexpectedly into a two-measure piano cadence, this time strongly reminiscent of the piano music of De Falla or Albéniz. With an overlapping upward trumpet glissando we find ourselves squarely back in the key of C minor (where the clarinets originally started) *except* that a totally foreign pitch, a softly held D-flat, emerges quite unexpectedly from the chord; and one measure later, as if by some musical slight-of-hand, we find ourselves in the key of D-flat major and in a traditional thirty-two bar song form, initiated by the solo guitar.

These seemingly opposite elements miraculously fuse into a totality, splendidly serving its function as an introduction to the main body of the movement. None of its ideas come out of the jazz tradition, strictly speaking, and only become "jazzified" by virtue of the jazz sonorities of the players and their subtle jazz-rhythmic inflection.

It is this fine line between straight jazz (in later movements typical Ellington and Basie passages appear) and various classical elements which Burns treads so well, and in so doing allows himself to swing easily to either side of the line without losing either the balance between these diverse elements or the central thrust of the piece.

The procedure is basically the same in all three movements: standard jazz forms surrounded by introductions, codas, and interludes which reach out beyond the confines of jazz, including some semi-improvised nontonal elaborations of the themes by the four-piece rhythm section of piano, guitar, bass, and drums. (These give a brief glimpse, incidentally, of what we find in a much-expanded format in the aforementioned Tristano quartet sides.)

Perhaps *Summer Sequence* never had the pervasive influence it should have had, and which lesser and more artificial, often bombastic compositions like those of George Handy for the Boyd Raeburn band did have. Perhaps Ralph Burns was both too subtle and too far ahead of his time. In any event, the work and its superb performance by the Woody Herman orchestra of 1946 gave young musicians of the time a brief but clear glimpse of how the boundaries of jazz could be stretched without any loss of identity.

BAND 2

The Clothed Woman
(DUKE ELLINGTON)

> Duke Ellington and His Orchestra: Harold Baker, trumpet; Johnny Hodges, alto saxophone; Harry Carney, baritone saxophone; Duke Ellington, piano; Junior Raglin, bass; Sonny Greer, drums. *Recorded December 30, 1947, in New York. Originally issued on Columbia* 38236 (*mx* #CO38671).

It is by now a cliché to say that Ellington was almost always ten years ahead of his contemporaries. In no work is this truer than in his remarkable "Clothed Woman" of 1947, substantially a piano solo with a few minor interjections from a quintet of supporting instruments. Cast in a simple ABA form, the outer sections are startling explorations of practices not then common in jazz: a freely atonal harmonic language and a commensurately free rhythmic/metric structure in the manner of a declamatory recitative. Jazz without a steady 4/4 beat was then, and still is today to some extent, a rarity; and only Thelonious Monk among the major early avant-gardists shared Ellington's interest in such rhythmic experiments. Harmonically and rhythmically the A sections of "Clothed Woman" could have come from the hands of composers like Szymanowski or early Schoenberg, but the inflections and rhythmic attack, the sense of "suspended time in motion" could only have come from a great jazz performer.

The diversity of musical styles, so much a part of the early avant-garde scene, is present in "Clothed Woman" too. The B section, a light, flighty ragtime interlude whose musical antecedents lay at least forty years in the past, provides a delightful contrast to the framing A sections.

BAND 3

Yesterdays
(JEROME KERN)

Lennie Tristano Quartet: Lennie Tristano, piano; Billy Bauer, guitar; Arnold Fishkin, bass; Harold Granowsky, drums. *Recorded March 14, 1949, in New York. Originally issued on Capitol 1224 (mx #3714).*

Frequently overlooked by jazz historians and writers, the early chamber-sized improvisations of Lennie Tristano of the mid- and late forties show us yet another of the different kinds of experimentation that were the result of the harmonic and formal breakthroughs unleashed by the bop movement. The twin banners under which this music presented itself were atonality and contrapuntal design. Although Tristano was able by 1949 (in such pieces as "Wow" and "Crosscurrent") to break away from a traditional 32-bar song format and its tonal base, in the earlier sides the improvisations were still anchored to a more traditional ground. Indeed, one of the fascinations of these performances lies in the way that the harmonic underpinning of Kern's "Yesterdays," for example, is stretched almost to the breaking point. But no matter how far afield Tristano and Billy Bauer may roam, they always return to home base, a process which gives a remarkable fluidity to the harmonic contours of the piece. The delicious harmonic/melodic collisions which occur throughout the performance, the result of its free contrapuntal-linear format, are not only among the enduring charms of these sides, but present an aspect of broadening the base of jazz improvisation explored by few other musicians of the period.

BAND 4

Mirage
(PETE RUGOLO)

Stan Kenton and His Orchestra: Buddy Childers, Maynard Ferguson, Shorty Rogers, Chico Alvarez, and Don Paladino, trumpets; Milt Bernhart, Harry Betts, Bob Fitzpatrick, Bill Russo, and Bart Varsalona, trombones; John Graas and Lloyd Otto, french horns; Gene Englund, tuba; Art Pepper, clarinet and alto saxophone; Bud Shank, flute and alto saxophone; Bob Cooper, tenor saxophone, oboe, and english horn; Bart Cardarell, tenor saxophone and bassoon; Bob Gioga, baritone saxophone and bass clarinet; George Kast, Jim Cathcart, Lew Elias, Earl Cornwell, Anthony Doria, Jim Holmes, Alex Law, Herbert Offner, Dave Schackne, and Carl Ottobrino, violins; Stan Harris, Leonard Selic, and Sam Singer, violas; Gregory Bemko, Zachary Bock, and Jack Wulfe, cellos; Stan Kenton, piano; Laurindo Almeida, guitar; Don Bagley, bass; Shelly Manne, drums and timpani. *Recorded February 3, 1950, in Los Angeles. Originally issued on Capitol 28002 (mx #5476).*

One of the musicians who broadened the base of jazz improvisation was Pete Rugolo, Stan Kenton's chief arranger and composer-in-residence in the late forties and early fifties. However, Rugolo never aspired to the absolute purity of contrapuntal design which so singularly motivated Tristano, but rather treated linear devices as only one of a larger arsenal of compositional techniques. "Mirage" (1950) is a striking example of how an expanded jazz orchestra, including strings and "classical" winds, could integrate such diverse musical concerns into a cohesive totality. Essentially it is what we call a pedal-point piece in which the harmonic/melodic continuity is spun out over a single pitch or ostinato bass, a device already thoroughly explored by composers like Mahler and Shostakovich, to whose influences "Mirage" owes a great deal. (The fact that the recurring four-note motive resembles the main theme of Wagner's *Tristan und Isolde* is, I think, pure coincidence.) But again, it is another example of how diverse stylistic and technical elements, some from separate worlds of music, could be welded together in a single work.

BAND 5

Eclipse
(CHARLES MINGUS)

Charles Mingus Octet, with Janet Thurlow: Janet Thurlow, vocal; Willia Dennis, trombone; Eddie Caine, alto saxophone and flute; Teo Macero, tenor saxophone; Danny Bank, baritone saxophone; Jackson Wiley, cello; John Lewis, piano; Charles Mingus, bass; Kenny Clarke, drums. *Recorded October 27, 1953, in New York. Originally issued on Debut EP 450 (mx #: none).*

Mingus's "Eclipse" performance espouses the same basic approach already exemplified in *Summer Sequence*—although I do not mean to imply that Mingus in any way emulates Burns—of saving the really advanced explorations for the introductions, interludes, and postludes, reserving a somewhat milder (more tonal) treatment for the song itself. Essentially Mingus's approach is contrapuntal or polyphonic, with the cello acting as a second voice to the solo vocal part, to a large extent letting vertical/harmonic relationships be the result of linear developments. It is one of many examples of what was then a growing concern to return to the earlier polyphonic concepts of New Orleans jazz, virtually forgotten in the swing era with the rise of the arranger and the incessant use of "block chord" homophonic writing. It also links up with various attempts to return jazz to a chamber music format, rather than an orchestral one, in which the individual instrumental voices function with a greater degree of linear/melodic independence.

SIDE TWO BAND 1

Egdon Heath
(BILL RUSSO)

Stan Kenton and His Orchestra: Buddy Childers, Vic Minichiello, Sam Noto, Stu Williamson, and Don Smith, trumpets; Bob Fitzpatrick, Frank Rosolino, Milt Gold, Joe Ciavardone, and George Roberts, trombones; Lee Konitz, Dave Schildkraut, and Charlie Mariano, alto saxophones; Bill Perkins and Mike Cicchetti, tenor saxophones; Tony Ferina, baritone saxophone; Stan Kenton, piano; Bob Lesher, guitar; Don Bagley, bass; Stan Levey, drums. *Recorded March 3, 1954, in Los Angeles. Originally issued on Capitol EAPZ-525 (mx #12449).*

The term Third Stream simply suggests the intermingling of two musical mainstreams, jazz and classical music, into one larger flow—though in recent years the pianist and composer Ran Blake has expanded Third Stream to include a multiplicity of other ethnic musics, an idea quite logical and inevitable in the American melting pot. In Third Stream, as elsewhere, the specific ways in which two (or more) musical traditions are combined or fused can vary tremendously. Thus, the combining has sometimes been done linearly—that is to say, in successive sections of a piece; or vertically—when disparate elements may be fused simultaneously, perhaps in concurrent layers or strands. Finally, there can be Third Stream pieces that represent a combination of both approaches, but in all instances the concept suggests an in-depth fusion of musical elements or techniques rather than a superficial *appliqué* or mere grafting of one technique onto another.

With this degree of latitude in the overall concept of Third Stream, it is inevitable that different composers will choose to emphasize different elements in different pieces. Thus, William Russo's "Egdon Heath" (Russo was at the time one of Stan Kenton's chief arranger-composers) eschews the typically explicit jazz beat as stated by a rhythm section, substituting various pedal-point or ostinato devices, and challenges the ears of jazz-oriented listeners with a free-ranging harmonic language which relates only tangentially to conventional tonality. In four sections, the saxophones provide ostinato background figures for the opening trombone solo (played by Bob Fitzpatrick); a pedal-point of cymbal rolls serves the same function for the brass gestures of the fast second part, which shifts in turn to a more conventional improvised alto saxophone solo (by Dave Schildkraut) and ends with a recapitulation of the opening section.

Editor's note: According to the composer the title of this work bears no relation to Gustav Holst's orchestral piece *Egdon Heath,* although both are inspired by a passage in Thomas Hardy's *The Return of the Native:* "A place perfectly accordant with man's nature—neither ghastly, hateful, nor ugly; neither commonplace, unmeaning, nor tame; but, like man, slighted and enduring; and withal singularly colossal and mysterious in its swarthy monotony."

BAND 2

Concerto for Billy the Kid
(GEORGE RUSSELL)

George Russell and His Smalltet: Art Farmer, trumpet; Hal McKusick, alto saxophone; Bill Evans, piano; Barry Galbraith, guitar; Milt Hinton, bass; Paul Motian, drums. *Recorded October 17, 1956, in New York. Originally issued on RCA Victor LPM 1372 (mx #G2JB7838).*

In "Concerto for Billy the Kid"—Billy being the young Bill Evans—the point of emphasis is the concerto form, adapted to Russell's "Lydian chromatic concept of tonal organization" and to the fact that the concerto soloist is essentially an improviser. The idea of fashioning a frame for a major jazz soloist had already been thoroughly explored by Duke Ellington in the late thirties in a series of four-minute "concertos" for members of his orhcestra. In Russell's brilliant expansion of that basic idea we can see how composition and improvisation are welded into a seamless totality, where both complement each other and operate in the same harmonic/melodic field. (An even more successful realization of these particular musical ideas was created a year later [1957] in the last movement of George Russell's classic "All About Rosie.")

BAND 3

Transformation
(GUNTHER SCHULLER)

Brandeis Jazz Festival Ensemble: Jimmy Knepper, trombone; Jimmy Buffington, french horn; John LaPorta, clarinet; Robert DiDomenica, flute; Manuel Zegler, bassoon; Hal McKusick, tenor saxophone; Teddy Charles, vibraphone; Margaret Ross, harp; Bill Evans, piano; Joe Benjamin, bass; Teddy Sommer, drums. *Recorded June 20, 1957, in New York. Originally issued on Columbia WL 127 (mx #CO58205).*

In my own "Transformation" a variety of musical concepts converge: twelve-tone technique, *Klangfarbenmelodie* (tone-color-melody), jazz improvisation (again Bill Evans is the soloist), and metric breaking up of the jazz beat. In regard to the latter, rhythmic asymmetry has been a staple of classical composers' techniques since the early part of the twentieth century (particularly in the music of Stravinsky and Varèse), but in jazz in the 1950s it was still an extremely rare occurrence. As the title suggests, the work begins as a straight twelve-tone piece, with the melody parceled out among an interlocking chain of tone colors, and is gradually transformed into a jazz piece by the subtle introduction of jazz-rhythmic elements. Jazz and improvisation take over, only to succumb to the reverse process: they are gradually swallowed up by a growing riff which then breaks up into smaller fragments, juxtaposing in constant alternation classical and jazz rhythms. Thus, the intention in this piece was never to fuse jazz and classical elements into a totally new alloy, but rather to present them initially in succession—in peaceful coexistence—and later, in close, more competitive juxtaposition.

BAND 4

Piazza Navona
(JOHN LEWIS)

John Lewis and His Orchestra: Melvin Broiles, Bernie Glow, Al Kiger, and Joe Wilder, trumpets; Dick Hixon and David Baker, trombones; Gunther Schuller, Al Richman, Ray Alonge, and John Barrows, french horns; Harvey Phillips, tuba; John Lewis, piano; George Duvivier, bass; Connie Kay, Drums. *Recorded February 15, 1960, in New York. Originally issued on Atlantic LP(SD) 1334 (mx #4254).*

In John Lewis's "Piazza Navona" (preceded by a brief fanfare) the emphasis is more on bringing the regal, stately gestures of later Renaissance music with all its massed brass-consort sonorities (as in the works of Giovanni Gabrieli) into relationship with Lewis's own brand of classical chamber improvisation (trumpeter Al Kiger is the other soloist). Another continuing interest of John Lewis's has been the relationship between the Italian *commedia dell'arte*, a semi-improvised street theater, and the improvisatory techniques of jazz.

BAND 5

Laura
(DAVID RAKSIN AND JOHNNY MERCER)

Jeanne Lee, vocal; Ran Blake, piano. *Recorded December 7, 1961, in New York. Originally issued on RCA Victor LPM(S) 2500 (mx M2PB5535).*

It is fitting that this album of Third Stream offerings should close with Ran Blake and Jeanne Lee's remarkable 1961 dissertation on David Raksin's "Laura"; for Blake is still, in 1977, the leading (and indefatigable) disciple of Third Stream doctrine. Having expanded it to include a broader ethnic and idiomatic base, he at the same time enjoys a national following which belies the frequent and recurring predictions of the demise of the Third Stream. The date of this cut—1961—is also significant because it coincides with the end of the first flowering of the Third Stream movement.

To fully appreciate the wide range of musical influences that motivate Ran Blake's music one must know some of his other works: a single piece cannot do him justice. On the other hand, Raksin's already highly chromatic "Laura" is the ideal vehicle for a Lee-Blake collaboration. Their extraordinary ears and their sensitivity allow them considerable latitude in searching out the deepest harmonic nooks and crannies of this standard tune; yet they always return to tonal home base—though in ways that can easily baffle the ordinary musician or listener. Here the many worlds of music—Schoenbergian atonality, Billie Holiday's sadly poignant laments, the American popular ballad, extemporization and composition—all intertwine and blend into a music that epitomizes the basic concept and highest ideals of the Third Stream philosophy.

21

Composing for Orchestra

This article, written in 1969, was originally published in an anthology The Orchestral Composer's Point of View, *edited by Robert Stephan Hines (University of Oklahoma Press, 1970). It was reprinted a year later in* Perspectives of New Music, *the advanced music journal.*

With a forty-measure excerpt from his 1965 Symphony *as a point of departure, Schuller discusses the kinds of performance and ensemble problems orchestra musicians and conductors may encounter in much of mid-twentieth-century music. The view represented therein—one most composers would certainly have shared at the time—should be seen in perspective, for unquestionably many of the problems to which Schuller alludes have in the intervening years been solved by most quality performers. Many of the compositional and structural techniques, seemingly still so formidable in the 1960s, have long been more or less assimilated into the performance mainstream.*

THERE IS MUCH DISCUSSION these days regarding the future of the symphony orchestra, the future—and indeed, the present—of orchestral music, and the place that both occupy in the view of today's composers. The most pessimistic prognoses predict the imminent demise of the symphony orchestra, or, at best, its maintenance as a mere museum custodian of eighteenth- and nineteenth-century European musical repertory. A corollary view claims that ever since the mid-1950's composers have been turning away from the orchestra for a host of practical and musical reasons. A few prominent conductors have even gone so far as to suggest that the young composers of today are no longer interested in writing for the orchestra and that the genre "symphonic music" has lost the attraction as a performance medium which it enjoyed earlier in the century.

Many of these dire prognostications are, of course, utter nonsense.

To a large extent, they merely reflect in various ways the confusions and inner apprehensions of those who make such statements. While there are undeniably serious hazards and problems attendant to the performance of new music, which affect composers, performers, and orchestra managements alike—some of these problems are brilliantly dissected elsewhere in this book—it is simply an irresponsible generalization to say that composers are no longer interested in the orchestra as a medium.

There are, to be sure, a handful of composers who actually prefer to express themselves in terms of chamber music or who, as a result of some discouraging orchestral experiences, may momentarily claim a total disaffection for the orchestra. There may even be some composers who genuinely believe that the orchestra—along with opera, another favorite subject of the doom prognosticators is dead. But surely the vast majority of composers would speedily accept an invitation or commission to write an orchestral work, not to mention the thousands of composers who, undaunted and unsolicited, have written orchestral works which for one reason or another languish on shelves or await the extraction of parts at the first sign of interest by a conductor.

This is not to deny that twentieth-century composition and its performance have changed in many fundamental ways. In recent years we have witnessed a serious widening of the gap between the composer's view of the orchestra and the actual performance capacities of most orchestral players and conductors. Ever since Schönberg's *Five Pieces for Orchestra* and certain passages in the late works of Mahler, composers have treated the orchestra more and more as a large "chamber music ensemble," thereby usurping the traditional structure and function of the symphony orchestra. This approach seems to have reached an ultimate stage in the works of composers like Penderecki and Xenakis, who frequently require even string players to perform completely individual and independent parts. It is a measure of the divergence between these two points of view that, although the use of the orchestra string section in this manner has been accepted for some ten to fifteen years by composers and conductors involved with new music, the vast majority of orchestral string players find this situation still completely novel, not to say suspect.

It is true that, given the problems in getting good performances of contemporary orchestral works, particularly complex or difficult ones, the young composer is apt to try his hand at chamber music on the simple premise that it is easier to gather together a woodwind quintet or even, let us say, a fourteen-piece chamber ensemble than an orchestra of eighty in order to organize a reading of his work. Lastly, it must be said that the orchestra, as a musical instrument—and probably as a social institution—will have to undergo some revitalization if it is to maintain itself artistically as well as economically.

It is inevitable that changes in the concept of the orchestra will originate largely with composers. Although it is sometimes difficult to see any order or pattern in some recent developments in contemporary music, it is clear that the extraordinary amount of experimentation already experienced in our century is beginning to wane and to coalesce into broader directions and schools of thought. Among these I see concepts of thinking and writing for the orchestra which, though new, have already demonstrated their viability and which are essentially within the reach of most enterprising orchestras and conductors.

The problem of artistic regeneration is not new, of course. It is certainly deplorable that many of our best orchestras are not capable of playing the works of Babbitt or Carter or Martino—not to mention late Webern—with anything approaching technical accuracy and stylistic conviction. But have we forgotten that Berg's *Wozzeck* took 137 orchestral rehearsals to prepare, and that a work like Stravinsky's *Le Sacre du Printemps* still presented enormous performance problems to an orchestra as fine as the New York Philharmonic nearly thirty years after the *Sacre* was first performed?

On the other hand, there is literally no orchestral work that I have yet seen—and I have seen many—that I would deem beyond the capacities of an ideal orchestra such as could easily be organized in the United States, given the appropriate financial support. This is tantamount to saying that there are no musical-technical problems in any score I have seen that inherently defy performance realization. That such an "ideal" orchestra does not exist is, of course, unfortunate, and is not in any way the fault of composers or their products. It is traceable, rather, to two primary causes: There is at present no conductor of any major orchestra who aspires to the formation of such an "ideal" orchestra, skilled in the performance of contemporary music; and the training and education of 90 percent of those musicians presently populating our major orchestras was virtually devoid of any contact with the performance, conceptual, and stylistic problems of contemporary music. Only among the younger musicians of today—and even then still too infrequently—does one find an interest in and an awareness of these issues.

If we consider that the age of today's (1969) orchestral musicians averages out to about forty years, it follows that their education—again taken at an average—took place in the 1940's, a period when the works of Webern or Ives, for example, were still virtually unknown and unperformed. Schools and teachers were totally unaware of such "radical" new music and still taught their students in terms of nineteenth-century concepts of harmony, melody, and rhythm. Only the exceptional talent penetrated the mysteries of such "esoteric" composers as Schönberg, Bartók, or Messiaen. That generation of players grew up with all the erroneous preconceptions and deficiencies regarding new music which are almost inevitable when a language changes as dramatically as the language of music has in our time. These changes have manifested themselves in many ways: the move from tonality to atonality, from symmetrically patterned rhythms to asymmetrical, irregular rhythms; from closed, predetermined forms to open-ended, individualized forms; from the simple one-to-one relationships of nineteenth-century melody and its harmonic accompaniments to the multiple, complex internal relationships of today's music, and so on.

The younger players of today, on the other hand, are frequently free of these pre- and misconceptions. Their musical education took place at a time when in many ways—through recordings, live performances, analytical books or articles—an acquaintance with new music was at least theoretically possible. For many of them, the language of atonality, of "irregular" rhythms, of disjunct continuity is the natural langauge of their time, a language with which they have no basic aural or perceptual problems. Many others, of course, though young in age, have been subjected to older concepts of teaching and thereby indoctrinated with the notion that new music is meaningless and its performance a fruitless endeavor, or in any event not legitimized by "audience" support. These young players, of course, inherit the prejudices and limitations of their teachers. Music schools and publishers of teaching material—with some notable exceptions—have not yet recognized twentieth-century music as relevant to serious musical activity, a fact which greatly impedes the acquisition of those technical and perceptual skills required for the performance of new music.

Nevertheless many young musicians have broken through the various assorted educational-bureaucratic barriers and by one means or another have acquired some knowledge of contemporary music, or at least an open attitude towards it. That they are at the present time still outnumbered and often intimidated in their pursuit of new music by their older colleagues is as inevitable as it is deplorable. (It is worth noting in this connection that many superbly equipped young musicians today are not aspiring to play in symphony orchestras at all. They feel that they can better preserve their love for music, their artistic integrity, and their vital interest in new concepts of composition by playing chamber music and free-lancing on a selective basis, even at the risk of considerable financial insecurity.)

The dialogue between composers and orchestral players has always been a precarious and intermittent one. At times it seems to have broken down completely. It is quite pointless for composers to blame this or that orchestral player for his inability to cope with certain contemporary performance problems. The fault lies more with our conductors and our music-educational system. The latter clearly fails to provide sufficient training and opportunity to deal with these problems, while the majority of conductors avoid the confrontation with new music altogether or pay mere lip service to it.

There is, for example, no acceptable reason why our music schools should continue to turn out musicians who have never been taught how to play a quintuplet of eighth notes *evenly* against two quarter beats; or for that matter, against three quarters (5:6) or four eighth notes (5:4); or how to differentiate precisely between 𝄿♪., [3: 𝄾♪𝄾] and [5: 𝄿♩] or similar rhythmic differentiations. There is no reasonable excuse for our schools' failure to teach our instrumentalists and singers that the release of a note is just as important as its attack. There is no intelligent basis on which our schools and teachers can maintain the notion that "melody" must move conjunctly and within a reasonably close intervallic range—ignoring all the while the countless examples of two- or three-octave melodic leaps in Mozart's works, to cite only one example. There is no justifiable reason why our young students should not be taught the new function of dynamic indications in contemporary music, that they are no longer merely decorative or expressive adjuncts of playing a note, but frequently define and delineate structural aspects of the work. There is no defensible rationale for the failure of many of our schools to teach ear-training which incorporates non-tonal intervallic hearing.

The list could be extended to alarming proportions. The fact is, however, that unless these issues can be effectively dealt with at the educational level, we can hardly expect to turn out professional musicians who will feel at ease in contemporary music. And inevitably the sheer accumulation of misconceptions spirals into a negative attitude on the part of musicians and conductors, which in turn transmits itself to the audience. The net result is that neither the audience nor the players establish sufficient rapport with the new music to acquire, in turn, the ability to discriminate between "good" and "bad" new music, between "significant innovation" and "faddish self-indulgence," between real talent and mere routine competence.

The conductor's role is just as critical as the "educator's," for he too must be, among other things, an educator—of the public and of the orchestra. An orchestra whose conductor cannot or will not consistently perform significant contemporary repertory can hardly become adept in it. An orchestra trained in a constant diet of Beethoven and Brahms gradually loses its mental, technical, and perceptual capacity to deal with anything but long-established musical ideas. Its intellectual and musical growth becomes stunted, and at the first encounter with a "new" rhythmic or stylistic problem, such musicians are apt to become defensive, automatically blaming the composer for their troubles and accusing *him* of unmusicality. The inevitable long-range by-product of this situation is the basic antagonism and mistrust between orchestral musicians and composers. This may not always become overt or harshly expressed, but the suspicion that all contemporary music is at worst the work of "fakes" and "frauds" and at best a nuisance or necessary evil lies just below the surface of most musicians' minds; and it doesn't take much to bring these feelings out into the open.

The conductor has it in his power to be the "mediator" between the composer and the musicians. He can bring to life for them—and for the audience as well—the lifeless blueprint which is a musical score. This he not only can do, but it ought to be his obligation. If he fails in this obligation, he not only thwarts the composer, who is after all the creative force keeping the art of music alive, but he deprives the musician of the opportunity to regenerate his intellectual and digital-technical capacities.

Human nature being as it is, it is inevitable that conductors—who are, I suppose, *human* beings—and musicians take the path of least resistance. Very few seem to identify with the sense of obligation, of responsibility, of dedication, and in turn of privilege that men like

Mitropoulos, Hans Rosbaud, or Koussevitzky brought to this issue, or that such younger musicians as Boulez, Charles Rosen, Paul Zukofsky, Bethany Beardslee—to mention but a few—bring to it today. In this connection, it is well to recognize that the "specialists" in contemporary music, regardless of their ability, perform only a limited service to contemporary music. It is much too easy for the traditional establishment to ignore these specialists as peripheral outsiders and indeed to point to the many among them whose musicianship is somewhat suspect, as is evident when it is put to the test in the classical repertoire. The problem under discussion here will never be alleviated until someone like a Karajan or a Szell puts his reputation and abilities on the line, so to speak, on behalf of contemporary music—not just on behalf of a few selected, isolated, predominantly "conservative" composers, but contemporary music in all of its variegated and truly contemporary manifestations.

I have perhaps been more fortunate than some in my contacts with orchestras and conductors, and I could not complain about the numerous excellent performances which some of my orchestral works have enjoyed. However, having also experienced on a number of occasions under-rehearsed, ill-prepared, misunderstood renditions of my works, I know all too well the feeling of helplessness and bitterness which the composer experiences as he hears his work massacred and senses the hostility of the audience rising about him. If I have enjoyed a more favorable relationship in this quarter, it is probably due in part to the fact that I was privileged to grow up within earshot of one of the great orchestras of the world, the New York Philharmonic, and subsequently spent the better part of my adult life in major orchestras as a professional instrumentalist. Viewing the orchestra, the conductor, the subtle psychological and artistic relationship that exists between an orchestra and conductor at close range and from the inside, as it were, I am sure provided me with insights into this special world which in turn must have affected my writing for the orchestra. I must also qualify my small success in penetrating the repertory of many world-renowned orchestras by saying that I do not claim to be the most "advanced" "radical" innovator. Such evaluations are extremely relative and change from year to year, and though many conductors regard my work as *avant-garde* and are frightened into inactivity by it, an objective viewpoint would place me more towards the center, albeit on the left side of the stylistic spectrum. It must also be added that within my accumulated orchestral works, it is consistently those which avoid the more interesting and complex orchestral problems of texture, rhythmic structure, form, and continuity that are performed. As a result those works which I consider to be my most important orchestral works are rarely performed, largely because conductors are frightened off by the problems they present, and it is therefore easier to turn once again to the *Seven Studies on Themes of Paul Klee*.

Having thus cast myself in the role of the "conservative radical," or the "radical conservative," it might be useful to touch upon some of the performance problems in one of my major orchestral compositions, which—though the problems therein are elementary to *me*—seem to cause such concern among conductors and orchestral players. It is curious that musicians are not intimidated by even the most extreme demands upon their instrumental *virtuosity*, but they are turned off by the slightest demands on their *intellectual* capacities. It has always been a disturbing curiosity of the orchestral player's mentality that he is challenged by the former (technical virtuosity) and demoralized by the latter (intellectual, i.e., mental virtuosity). Moreover, even this technical virtuosity must be presented (and demanded) by the composer in more or less familiar formats (i.e., conjunct lines, regular rhythmic patterns, traditional continuity, etc.) or else he "loses" the player.

In my *Symphony* (1965), commissioned by the Fine Arts Department of the Dallas Public Library for the Dallas Symphony, I was preoccupied with developing in my own way contemporary analogues to certain eighteenth- and nineteenth-century concepts which have been considered obsolete and unusuable in serial writing by many of the present-day style arbiters. Most particularly, I was interested in the symphonic form in the light of certain recently developed serial principles which permit the establishing of contemporary analogues for the basic precepts of diatonic tonality. The serial principles alluded to are those embodied in the concept of combinatoriality. I am assuming that this concept is familiar to the reader of this volume, or if not, he will turn to other sources for information on that subject. I will confine myself only to those external aspects of the work with which the conductor and musician must concern themselves, if they are to perform the work adequately.

Most performers would define performing as consisting of an accurate rendition of all pitch, rhythmic, dynamic, and timbral events as notated by the composer. Most composers would wish to expand that definition to include an understanding of the basic internal relationships (intervallic, rhythmic, structural, formal, timbral) in a given player's part which may not be apparent immediately or after a single reading of the work. Since every responsible composer I know in one way or another occupies himself with the delineation and clarification—for the purpose of greater expression and communication—of such internal relationships, which, in composite, form the external sonic surface of the piece, it follows that the performers must work towards a fully realized performance *from the inside*. That is to say, a merely surface rendition, no matter how technically accurate or skillful, will not suffice if it ignores the inner preceptual demands of the work. This is equal to saying that every player in an orchestra should understand in regard to every note he plays—isolated, disjointed, or not—with which other note (or notes) the first one is to be structurally associated. Only then can he play that note with the right sonority, right dynamic, right duration, right attack and release, and right *feeling*, so as to allow it to function properly in terms of the smaller and larger surrounding context, and finally, the grand design of the work. The ideal performance is one in which all the thousands of notes the composer has put down on paper after having carefully explored all the potential relationships radiating out from one note to all other notes in the work, are understood and performed in terms of those relationships. In some of today's more complex works, that is certainly a large order, but, as I have stated before, far from inherently impossible.

Of course, the musician does not have a score in front of him; he has only a single part and his immediate sight-reading concern is whether a given passage is a solo and exposed, or whether it is an incidental part hidden in the total instrumental fabric. Perhaps one day a clever inventor will discover a means by which orchestra players can play from a full score, as all intelligent string quartets do nowadays. In any event it is the conductor's job to elucidate for the performer the role and function of individual passages or notes. And he must keep at it until all the performers involved understand and *feel* the relationships elucidated.

The roles and functions which a passage or a note may perform are of an infinite variety, and even a single work may involve the player in a variety of types of structures. My *Symphony*, for example, contains several types of timbre melodies (*Klangfarbenmelodien*). Some are embodied in "free" homophonic forms (as in the first movement) or in "strict" polyphonic forms (as in the six-part fugue and four-part double-canon-by-inversion in the second movement). The second movement is indebted conceptually to both Bach and Webern, perhaps the two greatest contrapuntal masters of their respective eras. The opening fugue is organized in the manner of Bach's so-called permutational fugues, in which the various contrapuntal lines are combined in strict patterns of order (chosen, of course, by the composer). Each individual line in my fugue, however, is orchestrationally fragmented in terms of Schönberg's *Klangfarbenmelodie* (literally, tone-color-melody, in which a single melody is played not by one instrument but by many, linked together into a sort of "chain" of colors and timbres). This instrumental fugue is interlocked with a recapitulation thereof, this time in the percussion instruments, the entire passage serving as an interlude to the next section: the aforementioned double canon, which again makes use of "tone color melody." This too is recapitulated by the percussion, leading this time to one of the two possible alternatives to polyphony: monody. In complete contrast then, the solo horn sings a long-line melody, accompanied sparsely only at cadential points.

The fugue, although harmless looking in each individual part, is quite demanding in its multiplicity of linear relationships between parts and from one instrumental timbre segment to another. Example 1, shows measures 1-15 of the fugue in its simplest reduced form. Example 2, shows measures 9-15 of the same music in full score, showing the orchestrational fragmentation applied. The linear relationship from one instrument to another is indicated either by connecting dotted lines or by parentheses in the preceding and succeeding instruments for each contrapuntal line. Thus each of the six lines is "handed around" from instrument to instrument, as if played on some gigantic multi-timbered organ on which, by merely pulling a stop, different timbres can be instantly chosen for individual notes or groups of notes.

As can be readily seen from Example 2, the trumpet in measure 10 must know that he gets his note, as in a relay race, from the clarinet, and passes his line on to the oboe. The trumpet's *pp* F-sharp must match dynamically with the clarinet's three notes D, C-sharp, F-sharp. Likewise, the oboe in measure 11 must listen to the trumpet's crescendo in measure 10 and match its dynamic on the high C to the preceding F of the trumpet.

More difficult are the problems of attack and release. The trum-

EXAMPLE I

pet's attack on the F-sharp in measure 10 must be such as to continue the contrapuntal line coming from the first violins and the clarinet, and so as not to interrupt the flow of the line by either too soft or too hard an attack. Similarly the release of the dotted-quarter F must be precise so as not to leave a gap before the incoming oboe, nor may it overlap past the oboe's entrance. In addition, the exact release must be of such a nature as to musically pass the note on to the oboe. Beyond these points, of course, each line in its instrumental composites makes a single phrase unit, which in turn has to be balanced against the other contrapuntal lines. Even more problematic is the necessity to

Examples 1–6, used by permission of Associated Music Publishers, Inc.

EXAMPLE 2

C
D
Fl. 1 2
Picc.
Ob. 1
E. H.
Cl. 1
B. Cl.
Bn. 1
C. Bn.
Hn. 1
3
Tpt. 1
Trb. 1
3
Tuba
Hp.
Vln. 1
Vln. 2
Va.
Vc.
Cb.
9 Tutti con sord.
10
11
12
13
14
15
senza sord.
con sord.
senza sord. 4 soli
tutti
1 Solo
poco dim.
poco cresc.
espr.
a2

keep each line flowing, despite the fragmentation which allows each player to participate only in short segments of it. He must in other words feel the whole line, hear it along its entire course, insert his tiny segment, and pass it on to the next player, all the while feeling and hearing the over-all six-part progression of the fugue.

If one realizes that in the entire forty measures of this fugue there are nearly four hundred contact points which involve the precise matching of attacks, releases, and dynamics, one can gain an idea of the amount of rehearsal time required if, for instance, the orchestra is one that has never previously experienced this kind of orchestral writing. On the other hand, if the orchestra and each individual in it were aware of and experienced in this kind of orchestral technique by virtue of previous training and an occasional, though over a period of time consistent, encounter with such a problem, then the forty bars of this fugue could probably be sight-read nearly perfectly.

This, of course, brings us back to the point which I have already made in regard to education and training at the hands of our schools and teachers. If the musician encounters the particular technique under discussion here for the first time in, let us say, his third season as a professional in a major orchestra, when in fact this technique is already some sixty years old, then the education of that individual can only be considered inadequate. To further emphasize my point, I need only remind the reader that in this imaginary rehearsal of my *Symphony*, we have so far covered only 40 measures, whereas the entire work contains 401 measures, most of them with equally challenging problems to solve. Nevertheless, I maintain that with the afore-mentioned "dream" orchestra, the entire work could be played to perfection and much of it virtually sight-read.

In the double canon of measures 65 through 106 there is, in addition to the timbre-melody fragmentation already encountered in the fugue, the problem of certain doublings in the harp and piano. The purpose of these doublings is to extrapolate from the double canon and to articulate in the harp and piano two forms of the twelve-tone set. As can be seen in Example 3, the harp and piano sets (encircled notes) are in the same inversional and combinatorial relationship to each other as the main sets of the canon itself (see Example 4).

In order for these doublings to function properly, they must be played with rhythmic accuracy. As measures 77 through 79 in Example 5, show (these three measures are presented in the pitch canon of Example 6 by the encircled notes), the piano and harp double

EXAMPLE 3

EXAMPLE 4

P4	C♯	D	F♯	B♭	A	F	C	A♭	G	B	E♭	E		
I11		A♭	G	E♭	B	C	E	A	C♯	D	B♭	F♯	F	
I5				D	C♯	F♯	A	B♭	F	E	B	C	E♭	A♭	G	
P10					G	A♭	E♭	C	B	E	F	B♭	A	F♯	C♯	D

Harp	C♯	C	A♭	E	F	A	D	F♯	G	E♭	B	B♭
Piano	F♯	G	B	E♭	D	B♭	F	C♯	C	E	A♭	A

EXAMPLE 5

77
78
79
Horns
Tpt.
Trb.
Harp
Piano
1 Solo
Vln. 1
Others
Vln. 2
Va.
Vc.
con sord.
senza sord.
(2.Vln.)
(Vlns.)
(Tpt.)
(3.Hn.)
(Va.)
(I.Vln.)
cresc.
pizz.
arco
pont.
ord.

EXAMPLE 6

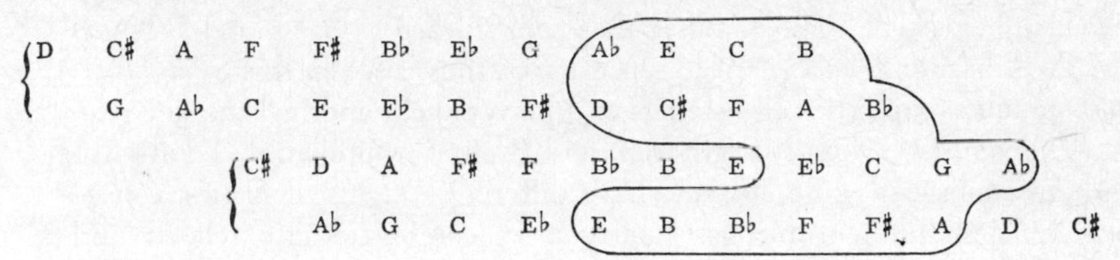

various pitches in the strings and brass. Since one of the canonic lines is in 9/8, the other in 3/4, the piano and harp participate in the main canon with both triple and duple subdivisions of the beat, very often in close proximity. If all players play these rhythms accurately—and they surely cannot be deemed particularly difficult or novel—the doublings will automatically occur. Since there is only one correct way of playing or one would think that there could not be any serious problems in achieving these doublings. But one would be amazed to see how many inaccurate variants of these rhythms players are capable of finding. Painstaking and time-consuming rehearsing is then the only answer.

In this three-bar segment (Ex. 5) there is the added problem of the rapidly changing dynamics in piano and harp. As can be seen in Examples 3 and 5, these result from the fact that the two instruments are doubling isolated notes from various contrapuntal lines all with their own dynamic progression (not serialized). Incidentally, in measures 78 and 79 the brass takes over the canon from the strings, momentarily causing other doublings which must be rhythmically meshed and dynamically balanced.

I have discussed only two sections of the work in which a great deal of intricate detail must be worked out in order for the performers to realize what I have written and intended. Other sections call for other kinds of relationships to be worked out: the durational-proportional relationships of the Scherzo movement, in which the ratios 4–2–2¼ are the primary determinants; the almost visual relationships in the first movement, in which an essentially vertical (chordal, non-melodic) music is gradually turned on its side, so to speak, to become an essentially horizontal (melodic, non-chordal) music. And finally there are the larger "key" relationships of the four movements of the *Symphony*, defined by the relationship of three different 'derived sets' in the second, third, and fourth movements to the prime set of the first movement. I believe that the time will come when the twelve-tone language will be so familiar that sophisticated players and listeners will hear and "feel" derived sets, such as those used in this *Symphony,* much as sophisticated listeners and players hear and feel the key relationships in Mozart, Beethoven, and Brahms symphonies. To a sensitive musician, different keys have different weights, densities, consistencies, different moods. Analogously, combinatorially related and derived sets reveal differentiating properties upon repeated hearings and with familiarity.

In that ideal future a work like my *Symphony*—and hundreds of equally worthy works by the many gifted composers of our time—will be performed accurately, with ease, with conviction, and with a sense of beauty which unfortunately is now reserved by our symphony orchestras and their conductors only for the nineteenth- and late eighteenth-century classics.

Music Performance and Contemporary Music

22

American Performance and New Music

Another stab at the complex problem of performing new music, primarily orchestral—again from the vantage point of the early 1960s. The article was written for the Spring 1963 issue of Perspectives of New Music. *It includes a by-now famous attack on some of the highly touted rhythmic-notational absurdities perpetrated by certain celebrated European composers.*

IN A COUNTRY as vast as ours, it is well-nigh impossible for any individual to assess standards of musical performance accurately, even as they bear upon an area so specialized as contemporary music. Musical activity takes place in such a large number of localities, is practiced by such a variety of organizations and on so many different planes, that one could probably make the available statistical data support any theory one wished to promulgate. In this welter of activity, however, one fact seems to stand out: the mean level of American performance standards has risen considerably in the last two decades at a rate that is proportionately higher than anywhere else in the world. Though the American composer may not always realize or appreciate it, he is the beneficiary of this widening spiral of improvement which applies to all styles and degrees of contemporaneity, and is informed and nurtured by the specific characteristics of our competitive social system.

This is not to say that there is no room for improvement on the part of performers, nor that performers have solved the practical problems occasioned by the greatly increased demands of today's music. But if we consider that Western musical tradition has undergone two rather drastic "revolutions" in our century (one in the first two decades and the other within the last fifteen years), the extent to which our musicians have been able to keep abreast of these rampaging developments is rather surprising.

Statistics relating to the performance of "contemporary music" such as those published by the American Symphony Orchestra League reveal that the greater part of this repertory is not "contemporary" in any sense that this much-abused term is understood in this journal. In such lists, anything composed in the twentieth century—even works such as Stravinsky's *Firebird*, and early Bartók and Prokofieff—are classified as "contemporary."

Obviously, statistics based on such premises are misleading and meaningless. Thus, within the context of this article, performance problems will be discussed in terms of that "contemporary music" which deals with techniques and concepts evolved in the last two decades, or with other aspects not yet assimilated into the mainstream of American performance practices—the music, in other words, which is often erroneously called "experimental" or "avant-garde."

While I am sometimes astounded at the excellence of performances of "advanced" music, especially (and most frequently) in the hands of the youngest generation of musicians, I am at other times left with the impression that too many performers, regardless of their technical competence, are totally unaware of the new problems that have been brought to the fore by recent compositional developments. This is even true—in fact perhaps most often true—of our technically best equipped professionals, who may have sufficient instrumental skill to play an advanced contemporary work immaculately from a mechanical point of view, but who often lack any rapport with the new concepts that may have inspired it.

All too often, conductors and instrumentalists still search for traditional harmonic functions, and are unable to accept the principle of intervallic autonomy found in most recent music. They are even less aware of the new roles played by such elements as dynamics and timbre, which have been accorded structural functions that can no longer be realized by means of the subjective, approximate approach of nineteenth-century interpretational mannerisms. (This, however, does not rule out the fact that in time a whole new set of "instinctive" performance practices, expressive of newer musical concepts, may be developed.) After nearly two centuries of that attitude, composers may have to exercise considerable patience until newer concepts are assimilated and the technical instrumental problems created thereby solved. Perhaps the first requirement for the realization of these goals is the recognition that new performance demands do not necessarily imply the discarding of the older criteria. It is not so much a matter of renouncing the old, although this is sometimes also necessary, as of extending and enriching our musical language by accepting the new. As Varèse once put it: "Just because there are other ways of getting there, you do not kill the horse."

The point-to-point continuity of contemporary music also remains largely misunderstood by performers. It is, of course, often simply rejected as the "aberration" of a "mathematically inclined" generation of composers. But even at a more sophisticated level, conductors and performers who have accepted the new formal and structural concepts as a *fait accompli*, often do not know how to bring them off. Consider, for example, the average musician's inability to cope with just two of the ideas most prevalent in today's new music: various uses of *Klangfarbenmelodie* and its extension into pointillism. In both, the single performer's previous role as an *individualistic* carrier of the melodic-expressive component is transformed into a *communal* one. The player no longer bears an entire melodic burden, but is asked to share it in specific ways with other instruments and players. Tone-color melody, in its most orthodox sense as a kind of "melodic dovetailing" distributed among several instruments, requires a degree of attack-and-release precision and timbral control that most players are totally unprepared for. To the player, the very idea that he is given only one note in a chain of melodic segments seems to be an affront to his carefully built-up, ego-nurtured sense of importance. It need hardly be emphasized that distortions in phrasing that are perpetrated in the name of "musical expression" are offensive enough in older music; but in a music predicated on a totally different kind of "expressive" content they are simply incongruous and absurd.

Some progress has been made in the realm of "attack" (or "touch") control, necessitated by the new structural and textural functions of this element. However, the manner of *release* is undoubtedly the most neglected and abused aspect of instrumental control. Whereas the romantic era taught us to taper notes gracefully, to mould them personally, and allowed for a certain degree of freedom in their release, contemporary forms of expression often require a degree of precision in bowing or breath control that approximates the very precise conditions of electronic music. This specific problem is very much related to the entire question of rhythmic accuracy, today's primary performance problem, which is discussed below. In "pointillistic" procedures, the instrumentalist's perplexities are compounded. Not only must he worry about the exact, often fractionalized duration of the note in question, but he must learn to play short notes with a refinement of sonority never before specified. Mostly as a result of misunderstanding, pointillistic or greatly fragmented structures often sound like a series of uncontrolled bleats and grunts. It has never occurred to some players that an isolated staccato note need not be choked off or blatant in its sonority (and consequently be unintentionally humorous). Both wind and string players would do well to learn that a short note is simply a shorter version of what could just as well have been a long note! The point is that the care in attack and tone control expended on a long note is equally required for a staccato note, if it is to form a meaningful moment within a larger continuity.

Questions of style in contemporary music also remain largely unresolved. While it is perhaps too early to expect stylistic differentiations in performances of the most recent works, it is precisely such qualities which make a composition memorable, and performers must quickly learn to direct their attention to them. Performances that capture this stylistic quintessence can do more to transmit an understanding of a new work than a merely pedantic rendering of its structural skeleton. In saying this I am not advocating a return to the "expressive" excesses of earlier periods, but simply remarking the necessity to strike a balance between the mechanical-technical aspects and the musical—the expressive—elements. Most often the latter can, of course, be deduced from the former, and in the case of the especially prescient musician, even intuitively felt. However, this link between the mechancial details of a composition and that which emerges between the lines, so to speak, in actual performance, is very rarely experienced in contemporary music; one tries in vain to call to memory more than a handful of performances that revealed the beauty and essence of the best of today's music. Similarly, one can think of a number of influential recordings of important contemporary works incorporating stylistic distortion (despite a degree of mechanical accuracy) whose effect will take many years to rectify.

I have already suggested that the most urgent problems in contemporary music relate to rhythm. Not only has variety of available rhythmic patterns increased manifoldly in our century, but the new degree of rhythmic-polyphonic independence requires a kind of accuracy for which the traditional limited repertory of simple divisive patterns leaves us ill-prepared. If one were to test a group of musicians of all levels of professionalism, with one extremely elementary rhythmic exercise, namely the playing of a moderately slow three against two, the percentage of accurate readings would be shockingly low. If we have not yet learned how to play a precise triplet, what then can we do with today's quintuplets, septuplets and other even more complex "irrational" rhythmic constellations?

But the performer is not alone in facing the challenge presented by newly acquired rhythmic possibilities. The composer has certain responsibilities too. Some new music abounds in rhythmic configurations that are literally unplayable. Whereas a few years ago composers were still legitimately involved in exploring new rhythmic possibilities, I doubt that any further *reasonable* rhythmic figures can be discovered, and it seems to me that composers today should concentrate on finding those complex or "irrational" rhythms that can be incorporated into a practical repertoire. If a composer must go beyond the point already reached, he is well advised to compose for electronic media. Many composers in their fascination with new rhythmic concepts seem to have "intellectualized" these beyond all performance realities; this is not less true because one or two inordinately gifted musicians (read: pianists) can approximate with a semblance of accuracy some of the still more involved rhythmic structures appearing in today's music. In ensembles or orchestras, such structures are merely absurd. Even if a passage such as the following example (hypothetical, but similar to existing ones)

EXAMPLE I

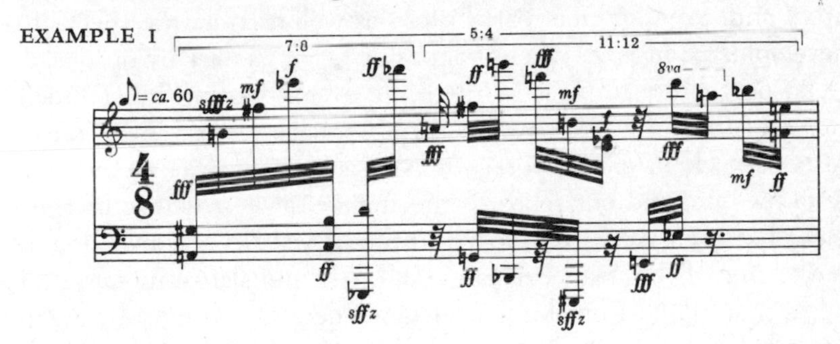

could be played accurately,[1] there are several more logical and practical notations possible. And beyond that, it is hardly likely that a

1. In Ex. 1 the first seven 32nd notes have a duration of 8/35 of a second per note; the remaining eleven, 12/55. Reduced to a more easily apprehensible approximation, the relationship of the two rhythmic patterns on a per-note basis is 4/17 to 4/18. I defy anyone to differentiate the two speeds accurately and with certainty in performance. Similarly, I find it impossible to believe that a musician could accurately compress five beats (or units) into the time of four, while *simultaneously* protracting two subdivisions thereof in two dissimilar segments, and both at different rates of protraction. The best that such a passage deserves is what it usually gets, an educated guess. (Ex. 1 is admittedly an extreme but by no means unrepresentative case.)

conversion into:

would result in a difference so vital that the loss in "serial" pedigree would not be more than outweighed by the increased playability of the passage. For any serial operations which have not been aurally-mentally and perhaps concretely tested by the composer are an aesthetic absurdity, and are bound to fail in terms of performance realities (at least of the "human" kind). It is one thing to sit for hours at a desk and devise the most complex rhythmic configurations that are notationally possible. It is entirely different to have to play them in split seconds.

I trust that these remarks will not be equated with musical reaction. Undoubtedly great strides will still be made toward enlarging the scope of rhythmic capabilities. But I would hazard the prediction that any such advances will be in the realm of increased sensitivity and accuracy and, to judge from past history as well as from the spectacular developments in jazz (where tempos of ♩ = 340 are by no means rare), in our ability to perceive rhythmic events passing by at much greater velocities. It is already demonstrable that the beat—in so far as it exists as such in contemporary music—is perceived at an increasingly faster rate, and our sense of the microcosmic structure of time divisions has been greatly developed in recent years. I seriously doubt, however, that the human ear can *accurately* translate into physical impulses that which is in the first instance determined only by arithmetic calculations (as in Ex. 1) or by notational mannerisms which actually represent ideas essentially external to music itself. For rhythms must ultimately be *felt* if they are ever to be played accurately. As far as *performance* goes they simply cannot remain at the level of intellectual apprehension, but must be translated into physical impulses. Reflexes can, of course, be trained to react to more complex impulses, but within each generation or each historical period there seem to be certain limits in this respect; at the very least, it takes a certain amount of time to acquire such new reflexive habits.

Many composers recognize that, with the recently developed extremes of complexity, we have also reached an impasse between notation and realization. They have therefore adopted new means of notation, which some have coupled with improvisational or "indeterminate" techniques that circumvent rhythmic problems in the above sense altogether. Aside from the aesthetic question involved in such concepts, they do not solve the "rhythmic" issue either, but simply evade it.

While rhythmic problems involve both performers and composers, questions of dynamics lie mainly in the hands of performers. Here again, there is a long-held tradition of bad habits, such as the ignoring of dynamics in all but the final stages of rehearsal, the absence of any precise definition of loudness degrees, and a general failure to understand their function. With today's tendency to abstract dynamics and give them a more functional role, their imprecise rendering is no longer merely a matter of carelessness or poor interpretation; now it subverts and annuls a vital structural element, in terms of which one no longer can speak of a "poor" performance, but rather of a "wrong" one. If the admonition of certain great musicians to "read the dynamics first" was sound advice in the past, it has become an essential today.

Another facet of the "dynamic" problem is the new way that sudden dynamic changes are used. At one time these were almost entirely a means of achieving contrast and "expressive" surprises. But today they also function as a means to delineate the internal design of polyphonic structures and larger formal schemes. The frequent use of highly differentiated dynamic levels not only requires maximum technical control of the instrument, but also a much increased mental agility. It would be difficult to say whether the musicians' inability to play a sequence of notes which has a different ungraduated dynamic for each note (our hypothetical example can again serve to illustrate an extreme case) stems more from lack of training or sheer resentment of such a passage. In either case, a re-evaluation of the functionality of dynamics in today's music (at least as a concept) is long overdue, especially since it would uncover the new-found beauty of dynamics used as a pure musical element that need not necessarily play a secondary role.

In the orchestra all these problems are aggravated, especially the rhythmic ones. Conductors—insofar as they are not themselves delinquent—spend the greater part of their rehearsal time correcting poorly read rhythms. Aside from the newness of the problem and the actual technical difficulties involved, there is one other fundamental reason for this dilatoriness, one peculiar to the United States. As a musical culture we are not only still young, but also not yet homogenous. That is to say, the individual traits of the different nationalities that make up our orchestras still persist. A musician with an Italian background "feels" rhythms quite differently from one with a German training. At this stage of our musical acculturation, the varied European influences are still to a large extent operative, and as a result very few orchestras can boast of a uniform rhythmic conception. Obviously a slight divergence at the most elementary rhythmic levels is bound to increase with more complex patterns. The extent to which the composer should take this factor into account is, I suppose, entirely dependent on his own musical attitudes. There are a few notable exceptions to this limitation, for example the Chicago and Cleveland Orchestras, which are in certain crucial respects the finest orchestras in America, having been trained by two of the world's foremost orcestral disciplinarians. It is therefore especially lamentable that they are very rarely given the opportunity to play any really contemporary music.

The conglomerate complexion of our orchestras has its good side too. Even an average American instrumental group plays with a natural rhythmic vitality and drive little known in Europe. The discipline of a good German orchestra is achieved at the expense of rhythmic drive and a kind of propulsive inner energy; their performances tend to be vertically accurate, but horizontally phlegmatic. No doubt this splendid characteristic of our American instrumentalists is an unintentional side-effect of the vitality of our popular cultures, particularly jazz. It also explains to some extent why much of the best American contemporary music is received with such indifference in Europe; when played by European orchestras, its innate characteristic vitality and virtuosity cannot be properly realized.

Improvement of this situation can only come if our educational institutions develop a closer awareness of the musical realities of our time. I think it was Mark Twain who said that we must cling to only that part of the past which can be of value to us in the present and the future. Considering the musical potential of this country, it is sad to think that we are best at producing pianists who win competitions in other lands playing Tchaikovsky and Rachmaninoff. Our genius and great talent may survive, but ultimately even they need the foundation and background of a culture which remains vital and in contact with the present, if they are to flourish.

23

Conducting Revisited

In this article, originally commissioned as a contribution to an anthology entitled The Conductor's Art, *edited by Carl Bamberger (McGraw-Hill, 1965), Schuller tackles some of the problems faced—or not faced—by conductors to whom contemporary music is an alien and enemy territory. It was one of only three "new" contributions to that anthology, the rest of this excellent volume—now, alas, out of print—consisting of various articles and statements by conductors from Berlioz, Wagner, von Bülow and Strauss to Furtwängler, Scherchen, Rudolf, and Steinberg.*

THE PROBLEMS relating to the conducting of contemporary music are, as might be expected, as varied and unpredictable as contemporary music itself. An unprecedented plethora of compositional schools, techniques, conceptions, and philosophies dominate the current scene, and it is, therefore, very difficult to generalize either about the compositional problems themselves or the performance and conductorial problems raised by them. The art of conducting is presently being subjected to some rather fundamental re-evaluations, and in a few instances new compositional approaches have radically changed conducting techniques or indeed eliminated them altogether.

However, leaving aside these extremes for the moment, if we compare conducting problems relating to contemporary music—which now, in 1964, ought no longer to include works composed in the first decades of our century—with those relating to eighteenth- and nineteenth-century music, we discover that there are certain crucial differences in conductorial requirements. These are in some respects differences in *kind,* but more frequently differences in *degrees.*

By that I mean that the conductor of a contemporary piece is trying to do essentially the same thing as the conductor of a nineteenth century work: to express the music with clarity, to shape it into that form which the composer indicates in the score, and to capture the essence of that composer's expression and style. Certain specific compositional techniques or styles will, of course, require rather specifically different conducting techniques, not to speak of a different musical orientation. But at the most fundamental level the conductor's job is still to provide a rhythmic frame of reference (through his beat) and a visual representation of the music's content (through the expression *in* his beat). On a purely technical level this would apply, for example, whether a given work made use of serial techniques or neo-classic or freely atonal principles of organization.

But what has changed in the performance of contemporary music is the *degree* of involvement and participation—I am tempted to say physical participation—required of the conductor.

In order to substantiate this point, we must first be very clear about certain fundamental differences between most twentieth-century music and the music of previous eras. A great deal has been said and written about the alleged increase of dissonance in contemporary music and the increased complexity of its rhythms, as if these were the only and primary problems for the listener and performer. But what really has complicated the performer's and listener's task are the radical changes in the *continuity* of most new music, i.e., its higher degree of variability and contrast. Not only does contemporary music involve a greater range of technical possibilities (instrumental registers, dynamic levels, timbral variety, rhythmic complexity, textural density, etc.), but the new forms of continuity involve a much greater *rate of change and contrast* in these respects. Whereas in earlier music it was highly unlikely that contrasts—other than in dynamic levels—would disrupt the even flow of a given phrase or theme (or indeed sometimes an entire movement), today one can almost expect the opposite. A single measure or a single musical idea may in itself involve a maximum degree of contrast in some or all of the above characteristics. And, of course, the progression from measure to measure or from one musical idea to another may also be marked by constant contrast and change. In a Mozart, Beethoven, or Brahms symphony we can reasonably expect a phrase to end more or less with the same instruments, the same number of instruments, more or less within the same range, and in the same meter and rhythm as it started.

In the music of our own time obviously no such guarantees can be made. Any one of a number of factors—singly or jointly—may serve to disrupt the continuity: changes of meter, of tempo, fragmentation of texture, use of extreme registers, of large intervals, of contrasting sonorities, and so on.

This new kind of continuity obviously requires a much higher degree of involvement on the part of the conductor. Purely statistically, there is more to control, more to shape, and at a greater rate of change. This does not yet take into account the fact that new music is, of necessity, less familiar to the performers, in terms of both specific compositions and general stylistic conceptions. Therefore on that account too, a contemporary work will require much more conductorial guidance.

It would appear that the points I have thus far raised should be self-evident. But unfortunately the majority of performances of new music offer no such evidence. Whether from a lack of knowledge of the specific composition involved or from a failure to understand these basic conducting requisites, too many conductors still conduct our new music as if it were shaped in eight-bar phrases and easy symmetrical patterns.

In this connection, it is time that several myths were disposed of once and for all. These center around the notion that the best conductor is one who uses a minimum of motion and physical energy. Any number of theories and half-truths are constantly invoked to perpetuate this idea; and conductors are fond of quoting stories associated, for example, with Richard Strauss, who by all accounts had an absolute phobia of sweating while on the podium, and who prided himself on the dryness of his armpits after a performance.

My point is not to question the quality of Strauss' conducting, which was indeed great, and altogether exceptional. Nor do I wish to imply that this approach is totally invalid. I am simply saying that in most instances it is not applicable in contemporary music.

It is not only entirely possible but highly desirable to conduct a Mozart symphony, for example, with a minimum of physical motion. Such a work to a large extent plays itself, and the conductor—once he has rehearsed any unsatisfactory details—merely guides and shapes the performance in its over-all form and expression. This is possible, obviously, because both the work and the style are familiar.

But if we move to a "contemporary" work by, let us say, Webern or Schoenberg or Babbitt or Nono, we are facing a totally different set of compositional and performance criteria. Performances of such works by conductors of the "no-sweat" school are with few exceptions disastrous. At worst they suffer from a complete lack of control over structural details, and at best from a lack of emotional involvement with the work. And in many cases this approach may simply be a suave camouflage for an inadequate knowledge of the score. I have also seen conductors, who pride themselves on their clean technique and a beat near the point of invisibility, resort of necessity to a large vigorous baton technique when faced with a contemporary piece. But at least these gentlemen allowed their innermost instincts—instincts of musical self-preservation, one might say—to supersede baton mannerisms and a bogus visual elegance.

It is, of course, not a question of either large or small beats, which is primarily a matter of conductors' personal styles. And I am certainly not advocating various forms of over-conducting (excessively large baton motions, incessant subdivision of the beat, etc.). I am simply suggesting that baton techniques must be related to the music they serve, and that contemporary music is apt to make entirely different demands in this regard.

It seems to me that the primary areas in which a greater conductorial control is required are those of "cues" and "dynamics." In music where continuity is characterized by fragmentation and multilinear polyphony each individual orchestral part consists by definition of short phrase fragments, sometimes—as in certain *Klangfarben* structures—even of single notes. Beyond that, orchestral writing of the twentieth century in general is characterized by a greater independence of each instrumental part, a chamber music conception first initiated in the works of Mahler and Schoenberg. All this, coupled with the player's unfamiliarity with new music, necessitates a much greater cueing ability and knowledge of the score on the conductor's part.

Similarly, highly contrasting dynamics—sometimes several in one measure, or indeed several dynamics simultaneously—must be clearly delineated by the conductor. In such music the average "common-denominator" beat, usable in earlier music, no longer can do the job. The scale of reference has simply been narrowed down. Whereas the conductor in earlier music might have had to change course in baton movements only at the beginning of a phrase or a thematic unit, he now—in extreme cases—may have to do so every beat.

In respect both to cues and dynamics, one might assume that, if a piece is sufficiently rehearsed—this is in itself usually a big if—the players might on their own initiative deliver all entrances and proper dynamics. In practice, however, even the most experienced orchestral players will play with more conviction and expressivity if their cues and dynamic levels are confirmed by the conductor's beat. The conductor, after all, is presumably the one performer who, by virtue of knowing the score, understands the multiple relationships of all the parts to each other, something the individual player usually does not. Nor can the player be expected to know this in unfamiliar music. Actually, orchestral musicians are trained to give what the conductor demands. But if the conductor demands nothing, the player—with very few exceptions—will take the path of least resistance, and give nothing beyond the most fatter-of-fact rendition of the notes on the page.

There are types of music which, once they are carefully rehearsed, can be left more or less to perform themselves. But even the most painstaking rehearsal procedures in contemporary music offer no guarantees for subsequent concert performances. Aside from the complexity of the musical relationships, the sheer energy and concentration required of the player in negotiating all the extremes of contrast and technical problems need constant substantiation on the part of the conductor. Anything less will result in the blandness and inaccuracies we usually get in performances of new music.

Another area in which conducting conceptions and techniques need to be forcused more precisely is style. Again the great proliferation of techniques and styles in our century is at the root of the problem. For I doubt if such stylistic opposites as Brahms and Wagner or Mozart and Beethoven require different conceptions of baton technique. But Schoenberg and middle-period Stravinsky do; and so does Webern and, beyond that, such distinctive composers as Boulez and Babbitt and Xenakis.

It is always difficult to divorce technique from conception, but the specific compositional techniques involved in the works of the three last-named composers and the conducting conceptions required to realize them are so different as to constitute differences in technique. Certain kinds of texture, continuity, and expressive content determine the specifics of these styles, and in turn require variegated conducting styles. This does not even take into account the radically different techniques required in the works of the aleatory and "indeterminacy" schools. Common to these is a beat which is no longer a *sine qua non* of conducting, either because the beat as an integrating unit has been supplanted by larger time sequences (such as intervals of fifteen seconds, for example, which are visualized by the "conductor" in the manner of the hands of a clock), or because conducting is assumed to mean the delineation not of beasts or metric units, but solely of actual sonic events (such as indicating only the initial attack of a sustained musical event and not the beats following it). This latter technique, which would seem to have far-reaching consequences since the conductor no longer conducts the rhythmic scaffold underlying a work, but instead the actual impulses that characterize the musical continuity—in other words, not the beat but the actual music—such a technique obviously places a much greater responsibility on the conductor. And it involves once again the reflexive capacities of conductor and performer, thus restoring a vital sense of spontaneity and urgency which, as a basic ingredient, is sorely lacking in much concert music.

To be sure, one aspect of conducting has *not* changed: the role played by the ear. The ear is still the final controlling arbiter, and in contemporary music, perhaps more than any other, no amount of baton dexterity can make up for deficiencies in either the ear or the mind. This is not to imply that one needs to know a Schoenberg score better than one by Mozart. It is simply to say that the Schoenberg score is apt to be more complex than the one by Mozart, and there is thus, purely statistically, more to learn and more detail to control. There is probably also a great deal of unfamiliar territory to explore, and—in turn—to transmit lucidly to the player. This cannot be done without an awareness of the compositional techniques involved, a thorough knowledge of the structural outlines which define the work, and a total immersion in the expressive essentials of that work's style.

It will be noted that this is in essence no more and no less than what was always required, and these qualities still mark the highest achievements in the art of conducting. But the demands made by the music have multiplied; and the conductor's art, if it is to continue to serve the music, must reflect these increased demands.

24

The Future of Opera

An article written for the June 10, 1967 issue of Opera News. *In it Schuller ruminates on the problems of contemporary opera, as seen especially against the background of experimentation in European (read German) opera houses during the 1960s. The problems articulated here are, alas, not much further toward solution.*

THERE IS MUCH TALK these days about the future of opera, what direction it will take—whether, in fact, it can survive as the distinct musico-dramatic form we associate with the opera masterpieces of the past. Some go so far as to suggest that opera is dead, or should be. What they probably mean is that "grand opera" is, or ought to be, dead. But then the question of what constitutes "grand opera" immediately presents itself. Indeed, any fruitful discussion of the subject ought to begin with a round of definition of terms. What *kind* of opera are the various factions talking about? In fact, has the word "opera" any specific meaning beyond that of a general collective term?

When young composers and theorists issue their proclamations regarding the imminent demise of opera, they are undoubtedly thinking of the *Rigolettos* and *Rosenkavaliers* of the past, and they are probably driven to their extreme position by the tenacious conservatism of the average opera audience. Indeed, as long as the majority of leading opera houses around the world adopt a museum policy with regard to repertory, one must admit there is cause for concern as to the future of opera. The conservatism of impresarios and directors, mixed with the economic hazards of producing opera, has precipitated on the one hand a crisis in modern opera and on the other the negativism of those who denounce opera. Many factors, musical as well as social, play a role, explaining much but solving little; the opposing positions are now so far apart that there is hardly any common ground on which to base an objective discussion. Both sides—the opera traditionalists and those who regard opera as obsolete—are carrying on stubborn, self-involved monologues, and a dispassionate dialogue seems less likely than ever.

Much of the problem in contemporary opera comes to us from the world of theater, and I believe many composers and critics make the mistake of confusing opera with theater. The theater, having revolted against the past by initiating the "theater of the absurd" and by stretching out into the world of "happenings" and mixed-media improvisations, tends to scowl impatiently at the traditionalism of opera. (We ought not to forget that opera is the granddaddy of mixed-media art forms.) In the non-art of a non-theater peopled by non-heroes, there is certainly no room for the romantic heroines, stances and attitudes of nineteenth-century opera.

Some modern theater (Beckett, Ionesco) is already operatic. Freed of the necessity for realism and representationalism, the modern playwright can treat words and phrases in terms of abstract, acoustical-musical material; he can build sentences into abstract, quasi-musical structures that have a life and meaning beyond the actual text. With this kind of theater, ask the anti-opera theorists, who needs opera? Opera originally turned to literature, marrying words to music; contemporary theater has transformed words *into* music, thereby making traditional opera obsolete.

But while the contemporary theater can nudge opera forward on certain points, it cannot transfer its own solutions wholesale to opera. Life in the arts is never so simple, so black-and-white. There are in the history of the arts many examples of cross-fertilization between one discipline and another. Poetry, painting, and music have all enjoyed periods of rapprochement, in which their individual techniques and conceptions have overlapped and cross-bred. One need think only of the symbolist poets and their relationship to music and composers. In the history of opera we see many points of view on the relationship between drama and music, ranging from the theater-set-to-music of Rossini's *Barbiere di Siviglia,* for example, to the quasi-symphonic non-theater of Wagner's *Ring des Nibelungen.* Verdi's whole development can be seen in terms of an increasingly deeper integration of drama and music; if we adhere to that particular definition of opera, Verdi is the ultimate opera composer. But there are many worlds of opera. Mozart's recitatives and arias are also opera at its highest, and who is to say that the "set piece" opera, revived in our century by

Brecht and Weill, is incapable of future development? The melodrama of Puccini and the expressionism of *Wozzeck* are other examples of what traditional opera can encompass. Stylistically, opera can absorb the entire gamut of contemporary compositional techniques, from serialism to jazz improvisations. And it may be that the future of opera will depend on its ability to absorb successfully these and other musical idioms.

There are those who, though not opposed to opera as such, cannot relate to the oversimplified, overromanticized attitudes displayed in most of the nineteenth-century staples of operatic literature. For them a partial answer might be the updating of staging, the psychological modernizing and embellishment of opera roles. There are dangers and limits here, some of which have been encountered by the Wagner grandsons at Bayreuth. But undoubtedly a more adventurous approach to staging the old masterpieces in terms of a twentieth-century perspective could, if intelligently and respectfully done, help stem the tide of disaffection for opera experienced by many potential audiences, especially the young.

The avant-gardists see all those directions as artistic dead ends and demand instead a kind of experimental theater, in which the various media of sight and sound will be abstracted and collated into a new entity, as yet unnamed. Some radical excursions in this direction have already been made, and we are now faced with the reality of "operas" in which opera singers are asked *not* to sing or act, in which music is replaced by all manner of noise and mechanically generated sound-backdrops, and in which plot has been abandoned in favor of theatrical ritual, non-action and non-meaning. Some of the subsidized opera houses of Europe have produced such experimental "operas," in which the pantomime and theatrical ritual of the extreme musical avant-garde is simply transferred to the opera stage. This kind of theater is producing its own kind of opera, say its protagonists. The question, however, of whether it has proved itself capable of replacing opera or surviving as a new stage form remains to be affirmatively answered.

This is all very well and may in the end produce a valid kind of visual-musical experience, be it a happening or a more organized composed event. But surely such manifestations, even if "successful," cannot completely replace the more conservative opera conventions any more than Italian *verismo* spelled the death of romantic opera. I'm afraid that, as so often with today's musical avant-garde, we are being asked to accept the theoretical blueprints before the actual works achieved have proved their capacity to survive. I would not deny the possibility of a major breakthrough in the radical direction indicated above, but at the same time I cannot see that the more traditional opera conventions are thereby automatically exhausted and doomed.

Many factors go into making a work artistically successful, radical conception being of itself no guarantee of quality. The real issues are fought on another, perhaps more complex front, namely that of the composer's (and librettist's) individual imagination, sensitivity, inventiveness, depth of vision, and the authority with which he can present these qualities. At that level, questions of stylistic pedigree become irrelevant; and on those terms an opera conception that might be called traditional from a particular historical perspective could still be infused with enduring, universal quality. Excellent examples of this are Stravinsky's *Rake's Progress* and Britten's *Peter Grimes,* works in which a clear dramatic conception, skill, originality, and artistic integrity are the ingredients that ensure communication with the audience. Berg's *Wozzeck,* though older than the *Rake* and *Grimes,* is couched in a more advanced musical language. But ultimately it makes its point at the same high artistic level through the integration of the above-mentioned qualities, not just by virtue of its radicality of conception. In other words, it succeeds in spite of its technical complexities rather than because of them.

An artistic form as complex as opera can make its impact in many ways. It may overwhelm us primarily through its dramatic power, the urgency of its subject; it may seduce us through the sublimity of its music; it may dazzle us with its visual splendor and stage spectacle; or it may combine all of these elements into a perfected balance. Last but not least, it may be the projection of these elements by a particular singer or cast that electrifies us. All of these ingredients, together or individually, may make great opera. And it is at this level and on these terms that opera, as it exists today, communicates to a relatively large and devoted audience.

To combine all these elements into a significant experience is not easy. And there are not many places, especially in the United States, where composers can gain the experience needed to write an artistically successful opera. For it takes—barring the exceptional genius—more than an intellectual comprehension of the various components of opera. It takes a *feeling* for dramatic situations, their pacing, the momentum required to carry an action successfully forward, an ability to set moods through music—in short, that mysterious ingredient that separates dramatic music from the symphonic, with the latter's potential for much greater abstraction of musical thought and technique. (Perhaps this is an area in which foundations ought to initiate a comprehensive apprenticeship program.) Then too, the voice is a very special instrument—though perhaps not so limited as some opera singers would have us think—whose unique properties and limitations must nevertheless be taken into consideration when writing an opera. This can be put in a brutally simple way: an opera, if it is going to survive, must be singable by a reasonably intelligent and capable cast of opera singers. Failing that, the work may still be on some other terms a significant musical work, but I doubt that it belongs in the opera house.

This brings us to what is probably the crux of the problem of opera as a contemporary art form. Opera is a very special musico-sociological phenomenon, with certain conditions and characteristics peculiar to it. Questions as to the future of opera will have to be answered, at least in part, on sociological terms. Impatient critics may wish to talk about changing these sociological conditions, but until such changes are made, it is naive to expect opera to change drastically or die. Opera is the most expensive musical luxury our society has devised. As such, it is inherently resistant to experimentation, at least of the more extreme sort. The kind of experimentation we can permit ourselves in chamber music, symphonic music, electronic music and present to a sophisticated, specialized audience of experts at contemporary music festivals and university centers, is not feasible in the opera house, even in the state-subsidized ones. The sociological and economic conditions of survival are too difficult to allow for mere experimentation of this sort. These conditions are more demanding than those of the theater, so that an analogy between opera and theater in this respect is not even relevant.

Perhaps it is this inherent unwieldiness and built-in conservatism (in a relative sense) of opera that prompts some people impatiently to await its abolition. It must be said at this point that in many ways it is easier to experiment in a radical way than to renew and build on handed-down traditions. And it seems to me that those who are most vociferous in claiming that opera should be abandoned as a relic of the past are precisely those who cannot cope with its creative demands. Rather than retreat in humility, they denounce opera and pronounce it dead. If they wish to change opera, they cannot demand this on purely polemical terms. They will have to write the work that will justify, in fact make inevitable, the change they advocate. Failing that, they had better keep quiet.

Perhaps they will some day have their wish fulfilled. But it is not a situation I can foresee, for there is much that opera can still absorb from recent musical developments and from the other arts without succumbing to attrition. Opera is still a superb arena for the projecting of dramatic situations, emotional states, philosophies, even political creeds, not to mention the sheer joy of experiencing that most personal of musical instruments, the human voice. There is a timelessness about these ingredients to which human hearts and minds will continue to respond. The care and feeding of opera is, of course, a precarious business; and if the operatic masterpieces of the twentieth century are still small in number, they have nevertheless proven that opera is capable of development and rejuvenation, and, like the Chinese of old, of absorbing its conquerors.

25

Toward a New Classicism?

At once a review of historical, technical, aesthetic developments in music since the beginning of our century and an assessment of where these "achievements" have brought us. And then a glimpse at a possible "way out." A lecture delivered at Goucher College in Baltimore on October 15, 1978.

BY NOW IT MUST SEEM obvious to everyone that what we call "contemporary music" (the music of the last few decades) has failed to capture the sustained interest of either lay audiences or professional performers; in fact, it has encountered a stone wall of resistance and apathy.

For years we ascribed this phenomenon to the ill will and incompetence of audiences on the one hand, and to the stubborn self-indulgence of contemporary composers on the other. As the participants chose sides, little passed between them aside from abuse. And as the gap between composer and audience has widened (or at least has maintained its rather considerable width), positions on either side have hardened, with many performers and listeners rejecting outright the new creations of composers, while many composers have turned from the audience or, worse, have openly declared that they are really writing only for themselves and each other.

Of course there are many well-intentioned attempts to close the gap, by supporting the work of this or that composer (and, by implication, of contemporary composers in general), and by attempting to bring a fresh and wider audience to this new music. There are thousands of festivals, symposia, conferences, all earnestly devoted to the cause of contemporary music and to the closing of that communications gap; thousands of societies, organizations, institutions and foundations exclusively devoted to propagating the species "composer" by diligently commissioning and performing new music, developing exchanges between the opposing sides, and in general trying to keep the lines of communication at least open, if not exactly humming with excitement.

Indeed, today's young composer has hardly ever had it so good. I need only think back to my younger years to recognize the dramatic difference between the number of life support systems to nourish composers in existence today as compared with then, the 1940s and 1950s. The young composers generally don't know or believe this, but it is a readily demonstrable fact.

Yet all these well-meant efforts have not ameliorated the problem significantly: audiences stand before our new musics baffled, irritated and uncomprehending, while the classics and many forms of contemporary popular music capture and hold the hearts and ears of those with whom we composers would dearly love to communicate. At one time we relied on the incoming generation as our Great Hope. "They will listen to our music, and understand us without the language barriers their elders had," we told ourselves. Alas, vain hope! For young people today are, if anything, even more conservative in their musical tastes than their elders. And, in a culture which feeds on the senses in a big way, particularly the visual and aural senses (films and music), they have turned to rock and other directly communicating contemporary forms for their musical nourishment. Rather than gaining young audiences, we composers have, by and large, lost that front line which artists of the past could be sure would come to their support, if only in reaction to the old.

This is a major problem and must concern us all for, obviously, an art cannot survive without an audience. The art of music cannot exist in a hothouse atmosphere, in a laboratory or a research station.

We composers should remember that after all the artistic, aesthetic battles have been fought and all injustices to misunderstood composers rectified, the audience—the large over-all audience, in short, the culture—is the final arbiter of that which survives. These societal verdicts can come early, or late, and at first may be confused and even misled. But sooner or later, as if with the aid of some marvelous artistic gyroscope, the ultimate assessment is made, and usually remains for a very long time.

I am a composer, and so I have thought about these things all of my life. But I have also been a fighter and activist for new music for many decades, as a performer, as a teacher, as a lecturer. Perhaps naïvely, I used to take great comfort in the notion that almost all new art, particularly if it is radically new, is at first rejected or greeted with apathy, and that it takes a generation or two for the audience to catch up with the front runners. By that time, of course, they are no longer front runners, but are either dead or well-ensconced in the pack, and have now been replaced by a new group of front runners. I resigned myself to the notion that the complexities of Schoenberg, Webern, and Ives would have to wait their thirty- to forty-year turn to be resolved and understood. The problem is that it is no longer thirty years; it's getting to be sixty years. And my earlier optimism has long ago been replaced by a growing discomfort that that old axiom really never had much substance. It was an illusion, a hope. Despite the best efforts of thousands of well-meaning people on both the producer and the consumer side, things have not changed significantly, and at best we seem to be in a sort of stalemate, with no one having quite the right pawns and queens and kings and bishops to move to a clear checkmate.

Even the scandals, the bitter feuds of the past, are no longer compelling. To recall with what vehemence those battles were fought, we need only think of the Schoenbergians and Stravinskians battling each other not so long ago (while a generally bemused audience sat by, uncomprehending, mostly not liking either). Such manifestations of struggle between the new and the old have lost their effect and have been replaced largely by apathy and rejection, deadly in their effect, because they are not even concerned with the issues.

It is time we all had a fresh look at ourselves, on both sides. Indeed this seems to be really the case, as a significant amount of soul-searching can be observed, not only in this country but elsewhere. And perhaps we are already moving to a "new classicism," which I suggest might be *one* way out of the present dilemma. *I* may speak of a "new classicism"; others are calling it "neo-romanticism." In Darmstadt, Germany—that erstwhile haven of radical avant-gardism—a new group of German composers has emerged, calling their style *"Die neue Einfachheit"* (the new simplicity). Still others, such as Harold Schoenberg of the *New York Times,* long a staunch adversary of all things contemporary in music, has hailed the new trend as a return of the "conservative" in music; and some would go so far as to relate this to a general trend toward conservatism in our society.

What then do I mean by the "new classicism"? Why, specifically, is it coming to the fore now and what specifically are its characteristics? I think in order to answer those questions we first have to look at the gains and losses during the last sixty to seventy years of developments in new music.

It cannot be denied that there were gains and that an impressive number of musical masterpieces has been created in these last six or seven decades. The earlier years of our century in particular contributed their fair share of musical monuments, innovative milestones in the development of the art, that have survived and remained in our repertory and in our consciousness. I'm taking as a threshold point those years, between 1908 and 1913, when in fact the language of music changed drastically; to put it somewhat simply, tonality went out, atonality came in, and all the elements of music (melody, harmony, rhythm, form, texture, and so on) underwent dramatic and seemingly irreversible changes. The alphabet hadn't changed, but the words and sentence structure had, and therefore the syntax and grammar as well.

What was gained in those years was, to put it somewhat all-embracingly, a new freedom—the freedom of choices, of mobility, of new forms, of new options, which included new things to do as well as *not* to do. That last is a crucial point, for it is as much what is not done as what is done that defines an activity or, in our context, an art. It is as much what one chooses to exclude as what one includes that determines the result.

The new freedoms gained in those crucial years before World War I embraced nothing less than whole new visions. Stravinsky in his *Rite of Spring* and Schoenberg in his opera *Erwartung,* Ives in his Fourth Symphony and even Debussy in his masterpiece *Jeux,* and Scriabin in his last Piano Sonatas and Preludes saw and heard something that no one had ever seen and heard before. The essence of that vision comprised not only new sounds and new forms, but a breakthrough which seemed to have limitless consequences, indeed which seemed—for the first time in hundreds of years—unrestricted, unconfined, and seemingly infinite.

This freedom, however, frightened composers and audiences so that before the 1920s had passed everyone had retreated into various forms of neo-classicism, returns to the past—Stravinsky to his brand of neo-classicism, Schoenberg to traditional forms, rhythm, and meters. Ives retreated into the insurance business. But before that general retreat took place, composers had pretty much dismantled the old hierarchical structure of musical elements, with melody and harmony in priority positions and rhythm, sonority, dynamics and other components at secondary or tertiary levels. Everything was now permissible; therefore everything was tried and dared. And so Debussy and Stravinsky and the young Bartók and Prokofiev wrote pieces (or parts of pieces) which were based almost exclusively on rhythmic or metric considerations. Varèse soon joined that trend. By 1908, Schoenberg had already written a piece that literally did away with melody and virtually even rhythm, leaving only harmony and timbre as the two main carrying elements of the music. The futurists and other *enfants terribles* of the period variously did away with harmony, consonance, conjunct continuity—some would add, with beauty and logic. The closed forms of the past were replaced by the open-ended forms of today.

Much of this was in reaction to what perhaps can be regarded as the excesses of late romanticism and some of it was indeed superficial, silly, and shortlived. But a lot of it stuck, and stuck so well that it is still with us. Irreparable damage had been done to the old system and although great works were being written—not only the ones I have mentioned, but other, later compositions like Stravinsky's *Symphony of Psalms,* Schoenberg's *Orchestra Variations,* Webern's *Symphony, Op.* 21, Varèse's works of the twenties, Hindemith's *Mathis der Maler* and many more—nevertheless, a basic rupture with tradition *had* taken place and things had changed, though whether for better or for worse was not quite clear.

A kind of stability developed in the twenties and held on through the forties, when unexpectedly a second revolution shook the world of music. This is that period which specifically precipitated the present crisis, and in which we first encounter the works of Babbitt, Boulez, Cage, Stockhausen, and Berio. The innovations of the earlier revolution were taken up once again and pushed to new extremes. Once again, with new media like electronic and computer systems coming into the field, vast new vistas were opened up, unheard of new freedoms were perceived, and virtually no controls or predeterminations were exercised—or at least the non-exercise of such controls became one of the real options, as witnessed particularly in the work and philosophy of John Cage.

I think we all regarded these changes and gains as positive, as marvelous new possibilities, as the unlimited joy of experimentation. Some composers exercised their freedoms with more responsibility than others, but there was almost no questioning of where we were going at such headlong speeds, and whether in fact we were looking at gains or losses.

It seems in retrospect that there were more losses than gains; and even the gains we thought we had made we didn't learn how to control and master. Many of these gains were more technical than substantive, more illusory than real, and we may have thrown out the baby *and the audience* with the bath water.

I feel that the problem goes much deeper than the changed specifics of the language: atonality, free rhythms and meters, disjunct melodies and themes, stream-of-consciousness forms, etc. It is a question of having been seduced into the pursuit of complexity and intellectualism for their own sakes.

Now complexity and intellectual qualities in and of themselves are not necessarily bad or destructive. Several hundred Bach fugues, a lot of Beethoven's last works, and some of the best music of our century are a clear proof of that. But it is finally a question of balance: how much of the one as against how much of the other. And if we can, with our sophisticated minds and sophisticated techniques and sophisticated new (electronic) instruments, push our music to the limits of comprehensibility, ought we not at the same time to balance that with a commensurate infusion of emotion, simplicity, comprehensibility, and humanism?

My point is that we have lost too much for the gains we have made, even the real, non-illusory gains. And we have done that because we have made the basic mistake of thinking that, in order to accept something new, we must totally reject something old. It was always an either/or situation. If you adopted atonality as your language, you had to reject tonality completely. The conservatives reasoned the same way, only in reverse: if you maintained tonality, you had to reject atonality. And so on it went through all the elements and components of our musical means. If you embraced asymmetric rhythms and meters, then you were somehow honor-bound to reject all symmetry.

There can be little doubt that we went through a very trying and difficult period, explosive, shattering, with an unprecedented number of ideas and concepts competing with each other in an overwhelming multiplicity and variety (again freedom without control). As a result in these hectic years we could not gain sufficient perspective on ourselves, and new experiments, new innovations would constantly throw us off our guard.

In trying constantly to keep up with the latest advances and innovations, and in trying to follow the dictates coming out of the various avant-garde centers, like Darmstadt and the ISCM festivals in Europe or some of the big universities here in this country, we never seemed to have the chance to sit back and look critically at our handiwork, and perhaps to begin to ask some tough, soul-searching fundamental questions.

Instead we went right on assuming that the postulates and premises set back around 1950 by certain composers and tastemakers were sacrosanct and unquestionable.

While we were gaining all these alleged new and exciting freedoms, we were also losing a great deal. For example, we lost the whole meaning and usage of melody and theme. With some exceptions, such as Britten and Shostakovitch, who were put aside as conservatives, almost no composer worked with a theme and with thematic and motivic development. And melody? That was relegated to the past; it was considered old-fashioned and subversive to the cause of modern music to indulge in a melody. Little by little the need for melody or theme was abrogated and it wasn't too long before composers no longer even thought of writing and using themes; meanwhile they lost the ability to do so—from sheer disuse. Writing melodies or themes, after all, may have been common to composers of the past, but it is a skill which does not come easily, requiring enormous talent and practice.

Another example of something we lost with nothing to replace it—and this is in the realm of harmony—is that wonderful mysterious thing that enables us in diatonic music to go from minor to major, or from major to minor. And how the great composers, like Mozart and Beethoven and Wagner and Mahler, used that phenomenon of our musical language. We lost the ability to deal in bright or dark harmonies—insofar as we thought about harmonies at all. We didn't need to lose that quality, but we somehow decided that it was no longer relevant to our new language, to our new concepts—it was considered expendable ballast.

There were other don'ts, as opposed to the all-too-few dos. You didn't, for a while, particularly in the fifties and sixties, repeat anything. Repetition was out, old-fashioned, much too obvious, crude, simplistic.

You also didn't use techniques like pedal-points. That was really a cheap trick. And you didn't use a recognizable form. The new line was that every piece now had its own indigenous form. Now, such individuality of form is a lofty, even awesome goal, which it is given very few composers to achieve in any significant way—maybe once in a lifetime. How pretentious and arrogant to think that we—thousands of compsers all over the world—could achieve such levels of individuality in respect to form in every new work—and still preserve intelligibility.

Perhaps the most dangerous "do," which we avoided like the plague was accessibility: relatively immediate intelligibility, memorability (recognizability)—all cornerstones of the musical tradition and of the great masterworks of that tradition until our own era. Suddenly it was shameful to write music somebody could remember, could immediately understand, or find accessible. If there was a more complex way of doing it, then that was inevitably the way it was done. In short, we lost simplicity, even though deep in our hearts we knew that in simplicity there can be strength.

But now, the worst "don't" of all: for all our talk of freedom, we had deprived ourselves of dozens of viable, exciting choices. We did not allow ourselves—a few exceptions notwithstanding—the choice to write a tonal chord, to mix tonality and atonality. We did not allow ourselves the choice to mix highly complex structures with simple ones. We did not allow ourselves the choice to write non-melodic, non-thematic music along with melodic or thematic sections of pieces.

We did not allow ourselves the choice of rhythmic complexities (that perhaps only the eye but not the ear could discern at first listening) along with rhythmically, metrically simpler continuities. We said it must be this or that, but never both. We had made of the exceptions—Schoenberg writing one movement of a piece without theme and melody, Stravinsky writing a totally rhythmicized music, Satie writing purposely a harmonically bland music, Ives superimposing abstracted mathematically conceived rhythms onto an otherwise traditional form—we had made of such exceptions the norm. We elevated these exceptional moments to the level of the very essence of our concepts and techniques. We sealed ourselves off in our radical "ivory tower" and then wondered why musicians and audiences—*good* musicians and *good* audiences—no longer followed us to these extreme ends.

Finally, there is something even more precious which we lost. If there is one characteristic common to the great masterpieces of our Western musical tradition, it is that they use and coordinate *all* the elements of music fully. And innovators in the past, be they Monteverdi, Beethoven, or Wagner—moved forward *on all fronts:* harmony, melody, rhythm, form, and so on. That fullness, that richness of experience we associate with the great music of the past, an experience in which all of our listening and feeling faculties are involved—that is something we are given only in the rarest of circumstances today. When have we had music that gave you goosepimples, that made you choke with emotion, that brought tears to your eyes?

When I propose a "new classicism" I do not mean neo-classicism as we once knew it, or a return to the past in terms of style and certain fashions of the past, but rather a turning back to those profound verities—and, dare I say, human truths—that are common to all great music, whether of the baroque or the romantic or the modern era. In other words, it is not so much a "return to" as an "analogy to." I dream of finding contemporary analogies to that glorious past from which we still have much to learn and which we should not merely discard. Nor is my new classicism merely a new form of conservatism. It is in fact the opposite: a daring confrontation with certain rather disturbing realities and a radical move to gain back much of what we lost. It is not so much about conserving the past, but rather about accessibility and communication with our lost audience. It is to bring the past *through renewal* to the present, to translate that which is eternal in the past into our contemporary terms—to find the contemporary analogy to those past cumulative traditions, not only by discarding and omitting, but by rediscovering and truly adding.

Thus far I have spoken primarily about what I view as the obligations and concerns of the composer. There is, however, another whole set of obligations on the audience side which is beyond the scope of this article, but which must at least be mentioned, lest the "blame" for our situation be construed to rest entirely on the composer's side. On the audience/consumer side the problem to be addressed lies in the realm of education, both of the formal kind as in schools and in universities, and of the informal kind—what I call the "environmental" education—which in our era occurs almost primarily through the media of television and radio. In order for the equation between the creator (composer) and the consumer (audience), actual and potential, to become balanced, we must establish musical literacy. By that I mean the ability to comprehend and enjoy a wide spectrum of musics just in the way that most people (at least in Western cultures) achieve verbal and reading literacy. We don't have to be a writer in order to read and enjoy a book. Similarly, we don't have to be a composer or a performer to enjoy a musical performance. What we must do is extend the literacy which most people have in regard to the older and traditional musical repertory to the newer works of our century. This is tantamount to acquiring literacy in a new language—in this case, a musical language described generally by such characteristics as atonality, asymmetric rhythms, more complex structures, etc.

There is no inherent and unalterable reason why such literacy could not be achieved. One does not have to be an "expert" to understand many forms of contemporary music, but one *does* have to be able to appreciate minimally the musical language in which this music is cast. Our educational systems, both formal and informal, being economically and politically determined by the population by and large, similarly make no effort to achieve the goals of which I speak. As a result, the composer of today finds himself totally isolated from the large public, from the public educational system, and operates in a kind of cultural "ghetto," or, worse than that, a vacuum.

So while I have talked much about the obligations of the composer toward his or her public, I want to emphasize that I see a whole other set of obligations on the part of the public and the audience *toward the composer*. If the public does not concern itself in a generally supportive, yet (one hopes) discriminating, way with the artistic creations of our own time, then we are in grave danger of having no musical creativity in the future. We will have no fresh music to listen to, and we will be relegated forever to listening to a musical museum of the past. No art form can survive if it looks only backwards. It must be able to look forward; and in order to do so, it must have not only the creators who are gifted enough to do that, but also an audience toward whom these creative gifts will be directed and communicated.

In any event, whether what is proposed here will come true I cannot say. I *can* say that many composers are having second thoughts about what we are alleged to have achieved in the last forty to sixty years. The concerns expressed here are not unique to me. But I do have faith that out of the ashes of this phoenix will rise a composer or a school of composers that will regain what some are now feeling may have been lost.

A "new classicism" could take many forms and speak in many musical tongues. My own personal view of such a renaissance—particularly as it is expressed in my own music—is bound to be but one among many. But in whatever form or shape, I see it not as a mere return to the past nor on the other hand a total abandonment of the skills and techniques we have acquired in the twentieth century, but rather a richer, more homogeneous balance of the old with the new, of the traditional with the experimental, of the expressive with the intellectual, of the need to communicate with the need to try the unheard, the unseen, the unproven.

26

The State of American Orchestras

This is perhaps Schuller's most famous—some would even say "infamous"—public statement; also one of his most misunderstood.

Originally given in the form of an address to the students at the Berkshire Music Center at Tanglewood at its 1979 Opening Exercises, the talk was expanded upon in an article for High Fidelity *(June 1980). Tackling the various problems faced by the tripartite structure of the symphony orchestra—musicians (and Music Director, when there actually is one), management, and trustees—Schuller pleads for a working together toward common goals, rather than the usual adversarial relationships, in order to survive in an otherwise inimical or apathetic social/cultural climate. Schuller's views are hardly theoretical, since his experience with orchestras is complete and of long standing: as a veteran performer (hornist) in orchestras for twenty-five years (the Metropolitan Opera Orchestra, New York Philharmonic, Cincinnati Symphony among others), as a conductor of virtually every major and minor orchestra in America and abroad, as an educator and trainer of orchestras (both at Tanglewood and the New England Conservatory of Music), and as a composer widely performed by orchestras all over the world.*

[Last June composer/conductor Gunther Schuller welcomed students to the Berkshire Music Center (better known as Tanglewood), where he is Artistic Director, with a speech that shook the classical music world. The former president of the New England Conservatory (and now president of the National Music Council) eschewed the usual innocuous generalities in favor of alerting the young musicians—mostly postgraduates headed toward symphonic careers—to the evils and pitfalls they would likely encounter in their professional lives.

Taking as his point of departure Tanglewood founder Serge Koussevitzky's dictum that as musicians "we must not use music, we must serve it," Schuller with "no great pleasure" warned that, in U.S. orchestras, joy "has gone out of the faces of many of our musicians. Apathy, cynicism, hatred of new music abound on all sides. Unbelievably, we have developed the art of reading [music] to such a high level of visual-technical competence that we are in imminent danger of no longer needing our ears except for the crudest of note repairs. We have accomplished the ultimate musical ingenuity (or is it indignity?): we have learned to transform musical performing into a reading, visual skill, eliminating the very thing for which music exists—hearing."

Schuller bemoaned the fact that the term "professional symphony musician" has begun to conjure up images of performers who are "embittered, disgruntled, bored, who have come to hate music. . . . As I travel around the country guest-conducting various orchestras, it is often former students, who once had that shine in their countenance when they heard or made music, who long since have lost that spiritual identification with music."

The blame for the "cancer" that has spread through all but a handful of American orchestras, Schuller insisted, falls on the unionization, or at least the union mentality, of our orchestras; on music directors, absent more often than not, whose ensembles play second fiddle to their careers; and on the consequent filling of the artistic vacuum by nonartists—boards and orchestra managements. "Consider that, in thousands of pages of musicians' union bylaws, federal as well as local, no matter what city, you will look in vain for any mention of the word 'art,' " he said.

These rules "concern themselves exclusively with money, with time, with durations of rehearsals and intermissions; in short, with ways of achieving an ever-increasing maximum of financial gain for an ever-decreasing minimum of effort and a minimum of product. Many American ensembles are now down to three rehearsals a week, regardless of the difficulty of the program.

"Or consider the spectacle of musicians getting up from their chairs and walking off the stage in the middle of a phrase—even in the middle of a note—because the clock has struck 4:30 or 5:00. How *dare* we interrupt music in such a brutal fashion? It is an insult to our calling, an indignity that we visit on the work of a master for whose sake we went into music in the first place, and whose genius may be greater than all the talents in such an orchestra put together."

Once a young extra horn player with the New York Philharmonic in the mid-1940s, Schuller recalled the players' resistance to the rare performances of Mahler symphonies, many of his fellow brass players assailing them "as an imposition and an unwanted evil. . . . Now as I talk to some of my colleagues around the country, particularly in the top ten orchestras, they are already bored with Mahler. From hate to boredom in forty years!"

"We used to think years ago that the cynicism of symphony musicians was explicable by reason of their status: as underpaid and held in low esteem as a profession by society. But now that many musicians' salaries are relatively respectable, . . . things are even worse than before. Indeed, the cynicism is worst in precisely those orchestras that enjoy the highest standard of living."

Recording? Television? Schuller called these "two of the remaining areas of pride left—and, of course, there is extra money for that too. But . . . notice when a given passage or solo is done, how quickly the instrument goes into the lap. No participation in the music, . . . no listening to what else is going on. . . . And the ears are functioning only enough to insure playing relatively on time and relatively in tune."

Schuller complained that orchestras have become too successful as businesses: "Their techniques of survival are now those of the American corporation, including the full panoply of managerial and public relations accoutrements," implying that selling a product is often more important than the quality of the product itself. He went on to decry the problems created by "absentee musical directors and orchestras run not by artists, but by committees.

"We must solve these problems before it is too late, for we have in the last fifty to sixty years created and built in this country—starting with the Stokowskis, the Koussevitzkys, the Reiners, the Toscaninis, and Mitropouloses, and their respective orchestras—the finest technical orchestral instruments in the world. But we are about to throw the baby out with the bath water if we don't address our energies and talents to the ills that threaten to compromise our art.

"In a world that certainly seems to have gone completely insane, the arts are becoming ever more precious. When I look around me, or when I contemplate the daily barrage of depressing news, I cling to the notion that perhaps music and the other arts represent a rare refuge for us all—if we can but preserve their purity."

We asked Schuller to expand on the subject for our readers.]

SINCE MAKING those remarks, I have been overwhelmed with declarations of support. Never in my professional life have I done anything that has elicited such a strong and positive reaction. And the reaction has come from *all* levels and categories of musical society, including those I had singled out and those directly related to the issues: board members, managers, rank-and-file musicians, union members and officials, and conductors. Apparently, what I articulated had been festering just below the surface for some time, waiting for someone to express it publicly.

Now all of this may come as a shock to the average concertgoer, who, from the perspective of the second balcony, sees only the "glamorous" side of concert life. That person might well ask: Why such dissatisfaction, such bitterness, such cynicism among musicians? And, of course, the answer brings us to the other side of the argument. Musicians can justifiably point to a whole catalogue of evils as the root causes of their plight, and one could easily write an article on that subject alone. Suffice it here to touch upon some of the major problems that have remained largely unaddressed.

There is, to begin with, the historical fact of the tyrannization of several generations of orchestral players by the great conductor-tyrants of the past: Toscanini, Reiner, Stokowski, and Szell and Leinsdorf in their earlier years, to name only the more famous (infamous?). They had absolute control over the lives of their musicians in human, social, and economic terms, and they consistently abused it. Small wonder that a generation or two of musicians developed a deep-rooted fear of and hatred for "the conductor." Inevitably, they, with help from their unions, rebelled against the many inequities of such dictatorial control.

The exploitation of musicians by tyrannical conductors is now largely a thing of the past. But old wounds have hardened into tough scars. The adversary relationship between musicians and conductors still persists, smoldering under the surface and liable to emerge at the slightest provocation, regardless of the conductor's disposition; he is simply "the enemy." In addition—and there is, alas, also much justification here—musicians regard most conductors as mediocre and unworthy of their respect, ranging from dull to downright incompetent, and likely to have been chosen for political or personal reasons rather than for musical, inspirational, or leadership abilities.

Symphony musicians have other "natural enemies" as well: the management and the board of trustees. The former generally carries out the policy and economic directives of the latter, but local variations in the relationship between the two may make for varying degrees of responsibility. In most negotiations, the management finds itself more or less automatically, though not always wisely, on the side of the board. But the most crucial "blame" often falls to the board alone, for its choice of the conductor or music director. As the musicians see it, the person with whom they have to spend 90 percent of their professional lives is chosen by lay people who often have no idea how to select the "right" conductor. Musicians argue, therefore, that it is the incompetence of trustees and management in making decisions, compounded by a whole series of inherited inequities, that has caused them to rise up aggressively against their "tormentors," seeking decent living conditions and some influence on or equity in the major decisions that affect their lives and their futures.

A study of both history and human nature teaches that evil on one side usually begets evil on the other: thus the never-ending cycle of action, reaction, and counterreaction that has characterized labor-management relations throughout the industrial age. The tragedy of this process is that it institutionalizes the adversary relationship, thereby lessening the hope of breaking the cycle. The wounds on both sides become deeper, and each reaction arises out of an increasing well of bitterness, resentment, and frustration. All of this sociological paraphernalia is particularly inappropriate (and it seems to me unnecessary) in the arts.

If the symphony orchestra is to survive, musicians, management, and the trustees must collaborate seriously on their *collective* future, must develop a respectful, serious, substantive dialogue rather than yell at each other from entrenched positions. They must work out a common future based on a common process (and progress) that will deal realistically with the real "common enemy": those millions of people in our society—and their political representatives—in whose lives "classical" music plays no part whatever. *That* is the real dilemma! And as long as the three parties continue to beat each other over the head in a kind of fratricide, the old dictum "divide and conquer" threatens to prevail.

The sorry annual spectacle of orchestras on strike, with all the attendant bad publicity and generally woeful misrepresentation in the press, has to stop! We in the arts are altogether too vulnerable in this society to kill each other off year after year and hope to survive in the long run. As long as unions, orchestra committees, and members of ICSOM (the International Conference of Symphony and Opera Musicians—the voice of the classical orchestra player within the musicians' union) go to battle only for the immediate, short-term goals of better pay for fewer hours, shorter rehearsals, and correction of other real or perceived inequities without also considering the long-range questions of artistic progress, of how to generate greater support for their "product," of how to effect more widespread musical literacy in our society, no meaningful progress can be made on the real problems.

The achievements and accomplishments of ICSOM over the years have been well noted and appreciated, and it is clear that at times tough, aggressive tactics are the only approach that will work with recalcitrant managements and apathetic boards. But insofar as the ICSOM approach has concerned itself almost entirely with short-term and material improvements, mostly of a financial or "working condition" nature, it has fallen short of the mark. There are higher and better goals to which it, along with the unions, might also direct itself.

Most symphony musicians regard such goals as none of their concern. For example, some years ago, musicians and unions launched a drive for full employment and a fifty-two-week year, not in itself a reprehensible goal—indeed, a perfectly normal aspiration. Many orchestras got what they wanted. The only trouble was that they had given no thought to the harrowing question of *how* the managements were to fill those fifty-two-week seasons: Where would the concerts, the fees, the income to support that policy come from? All too typically, the musicians' position was: "Let management figure it out. That's not *our* concern."

Some are beginning to see now that it *was* their concern, or should have been. What managements were forced to do to honor those demands was to grab at any kind of concert or gig that came along. The so-called "runout concerts"—sometimes three a day in acoustically atrocious high school auditoriums or shopping centers, preceded and followed by exhausting bus trips, the whole venture often artificially generated by state or federal subsidies—have become the bane of the musicians' and managements' existence. They are most often an indignity to the profession, let alone the art. (That such concerts often provide the first contact with classical music for people in outlying areas is, of course, a positive factor. But at what price of wear and tear on the orchestras and their morale?)

I have taken musicians to task for their cynicism and apathy, but if pushed to name the one party most responsible for our present dilemmas, I would have to finger the boards. Not that there is anything inherently wrong with the American system of a lay board working with professional management and personnel; it's just that the system has been woefully abused. Out of sheer ignorance or arrogance, many boards have time and again made lamentable decisions: bad choices of conductors; stubborn refusals to accommodate the professional dignity of musicians in terms of salary, working conditions, or social benefits; failure to acknowledge the real problems. And too often, they carry out their functions in an elitist, nineteenth-century, social-tea-amidst-potted-palms manner.

One of the more disconcerting examples of board behavior occurred recently, when Michael Gielen was chosen as the new "musical director" of the Cincinnati Symphony Orchestra. Gielen is a fine, intelligent conductor, but the method by which the choice was made was outrageous. The selection was made by two or three trustees after seeing Gielen at only one concert—and that with the Detroit Symphony. He had never appeared in front of the Cincinnati Symphony and thus was totally unknown to the orchestra, most of the trustees, and the Cincinnati public. Not only was this the kind of affront that musicians quite understandably might resent, but it put the fate of an orchestra and the cultural welfare of an entire city in the hands of a couple of not very well informed citizens. In addition, it is clear that almost no consideration was given to American conductors, a theme so effectively investigated by John Rockwell in two recent *New York Times* articles.

What is most strenuously needed is the *education of trustees*. Wealth, a position of power and influence, and a vague love of music as an "entertainment" or a "social grace" can no longer be the major criteria for board membership. We should have obligatory training courses for prospective trustees, every bit as tough and demanding as those to which musicians are subjected.

One of the most serious problems facing the modern symphony is that of the absentee music director. Jet travel has afforded the possibility of simultaneous directorships, and conductors have shamelessly indulged themselves, taking on as many as three or four orchestras and, in one case some years ago, even five. Since boards have generally not been wise or strong enough to resist such temptations, many orchestras have become directionless, amorphous aggregations with no personality, style, or point of view. Their season is often a mere stringing together of programs—usually of the classical hit-parade variety—that permits no artistic growth of either the orchestra or the conductor, much less of the two *together*.

Time was when such growth—the orchestra learning from its conductor and vice versa—was held in high esteem. The subtle process of feeding off one another artistically is absolutely essential; no great orchestra was ever developed without such cross-fertilization. But such symbiotic relationships cannot develop overnight or during the infrequent intervals between a conductor's lengthy absences.

Many orchestra musicians are content with the annual round-robin of guest conductors and without specific directorship. But they are dead wrong. In the long run such an approach breeds complacency and an uncritical attitude. And it leads to confusion, since there are no particular performance criteria to aspire to, or to the development of a facile skill in following all conductorial comers regardless of quality. At worst, it leads to a kind of group arrogance, with the orchestra deciding how things shall be played, in default of genuine leadership. And it leads, finally, to an uninvolved, businesslike attitude: "Just let me play the notes, and don't bother me with stylistic niceties and all that artistic crap."

This philosophy of non-leadership and non-responsibility, usually promoted by precisely those musicians who have become most cynical and apathetic—often unfortunately the real leadership of an orchestra—is in turn welcomed by many managements and boards, since it makes the job of finding and keeping a "music director" that much easier.

Ideally, a music directorship is a full-time commitment, not only to the orchestra, but to the community, one that cannot be measured necessarily by how many weeks a conductor is actually on the podium. In some of the major orchestras, music directors conduct as few as eleven or twelve out of thirty weeks, in some a few more, and in some rare cases slightly more than half the season. Whether the conductor knows it or not, and likes it or not, he is the prime music educator of the community. Through his programs and artistic decisions, his musical and extramusical (including financial) concern for the orchestra, his commitment to the orchestra as an institution and the building of its audience—or through his *default* in these matters—he either educates well or educates poorly. If the latter, he is not a good music director, even though he may be a good conductor; he only takes and doesn't give.

Some conductors argue that, with the growing power of orchestra committees, they cannot be true music directors anymore, but become mere figureheads. Hiring and firing are largely in the hands of the ensemble itself, and programs and soloists are determined or heavily influenced by management to suit the box office. A board will quickly descend on any conductor who harbors notions of adventurous programming. These conductors ask, perhaps rightly: "What is there for me to be music director *of?*"

With the trend toward absentee music directors, another deplorable trend developed: the placing of more and more artistic/musical decision-making in the hands of the managements. *Artistic* decisions should not be made by *non-artists*. Period. They can be arrived at in collaboration and consultation with non-artists, but not made solely by them.

This sorry trend may have run its course, however. While two of the few remaining great music directors vacated their posts just last season—Ormandy in Philadelphia and Abravanel in Utah—Giulini seems determined to take on a real directorship, with all its attendant responsibilities, in Los Angeles. And elsewhere, there seems to be a growing realization that the indulgence of conductors and their Fifty-seventh Street agents has gone on long enough, that artistic direction and administrative management are not interchangeable.

It would not be difficult to go on citing other orchestral problems and inequities, large and small. But the implied questions remain: Why must these inequities, real or imaginary, lead to cynicism and apathy? Are bitterness and divisiveness the only possible reactions to these problems? And even more profound: Why do musicians let the very thing to which they have dedicated their lives—music-making—become the object of their hatred and cynicism? Why must one loathe the making of music and music itself?

One clear sign that the malaise infecting our symphonies is no deep dark secret is that, in increasing numbers, young musicians are turning away from the orchestra as a career. They pursue careers in chamber music, teaching, free-lancing, or even outside the music profession rather than face a life of regimentation and apathy in a typical symphony orchestra.

It needn't be so, of course. There are *many* orchestral musicians who do not let their art, or their love for it, be negatively affected. Though such musicians are not often loudly vocal and not, as a group, aggressive or belligerent, neither are they naïve or cowardly. They simply have the ability to rise above the sometimes petty issues. While they, too, may fight for better conditions, they will not let that become their sole or overriding motivation.

In any case, it is not a question of one group being pitted against another. What is needed is working together. The symphony orchestra is too fragile and vulnerable an institution in a basically apathetic society and a hostile economic environment to survive internal divisiveness much longer.

There are some positive examples of superb music directorships. Take the case of Maurice Abravanel, whose tenure of more than thirty years as conductor and music director of the Utah Symphony ended just last season. Abravanel developed the Utah Symphony into not only a first-rate orchestra, but one with exceptionally high morale and a very positive attitude and work ethic. He maintained a point of view: about the orchestra as a musical instrument, about repertory, about the place and function of the orchestra in both the musical and the general community. As a result the Utah Symphony developed a style and a distinct personality that it and the community could be proud of. Salt Lake City could identify with it as most places identify with their sports teams. One tangible bonus was a long-term recording contract that brought the highly praised recordings of the Utah Symphony to the rest of the world.

But Abravanel went further. Fully realizing his role as an "educator" of the orchestra, the audience, and the greater community, he developed excellent relationships over many years with the state legislature and with city officials, seeking their support—and getting it. And as a kind of fallout of his efforts, support began to come in for all kinds of other arts organizations—theater, ballet, museums, education in the arts—so that Utah and Salt Lake City have some of the finest performing groups and arts institutions anywhere in the United States.

What Abravanel achieved should be a lesson to us all. But such things take time to achieve; and time is precisely what few people are willing to give to anything anymore. Artists and musicians and conductors want instant fame, instant careers, instant success, without putting in the time and effort that such results require.

27

Program Notes—Various

Herewith a brief selection of diverse program notes supplied by Schuller for his own compositions on the occasion of their world premieres or subsequent performances. Chosen from a total of 110 works to date (1985), these four samplings display some of the diversity of Schuller's oeuvre, *although readers may find in them some recurrent themes.*

Concerto for Contrabassoon and Orchestra (1978)

I had had the idea of writing such a work for a long time, inspired by such wonderful passages for the contrabassoon as are found in Strauss's *Salome,* Glière's *Ilya Mourometz Symphony,* Ravel's *Mother Goose Suite,* and two of the great wind *Serenades*—the one in D minor by Dvorák and Mozart's *Gran Partita in B-flat*. But even there, with the exception of the two *Serenades,* the contrabassoon is generally stereotyped as an instrument limited to the depiction of evil, of monsters, beasts, etc. (the depraved and twisted *Salome,* the brigand Solovei in *Ilya Mourometz,* the Beast as opposed to the Beauty in Ravel's Suite, and other assorted misfits and monsters in the romantic literature). Indeed, one can measure the contrabassoon's standing among orchestral instruments in proportion to its close relative the bassoon, which is still perceived as the "buffoon" of the orchestra. The contrabassoon fares even less well, alas, and locker room jokes, mostly unprintable, and snide asides that the instrument's name really means "against the bassoon" abound.

All of this is, I think, quite unfair, and says more about our own lack of perception than about the actual character or capacity of the instrument. An instrument is what a composer and a performer make of it . . .

There is another aspect of composing a concerto for contrabassoon that I would like to mention. Since 99.5 percent of the world's musical ears quite naturally expect a melody or theme to occur in the upper or middle range, the composer of a contrabassoon concerto faces the special compositional/technical/acoustical problem of attracting the listener's attention to the nether regions of the human auditory range. (The highest note at present attainable on the contrabassoon is C-sharp directly above middle C.) The simplest way to do this would be to let the contrabassoon play unaccompanied and without any distracting interference from other, more 'normally' voiced instruments. But such a work would quite probably be a bore, and in any case not a concerto. Thus I faced to an unprecedented extent the unusual problem of establishing for the listener the unquestionable soloistic priority of the contrabassoon, i.e., of consistently attracting the listener's ears to that lowest range of our auditory spectrum, lest mere accompaniment or secondary passages—let us say, in the flute or violin—might be heard as primary, while the real primary material in the contrabassoon might be ignored or perceived as secondary. Tactful disposition of such elements as dynamic balance, density, degree of activity, and such old questions as the relationship between melody (or theme) and accompaniment became extremely crucial in the very composing of the piece.

The *Concerto* is in four movements, the third and fourth being linked by a solo *cadenza* and played without interruption. The first movement begins by contrasting the contrabassoon in its lowest register with the high violins, piccolos, and a sprinkling of celesta, harp, and glockenspiel. This more-than-five-octave registral gap is gradually filled in by the addition of other instruments (clarinets, bassoons, violas) and the gradual ascent of the contrabassoon into its highest range. All participants meet at the logical rendezvous point—middle C. A second subject puts the contrabassoon through some challenging technical paces—leaping passages, careening runs and the like—only to recapitulate the opening section, now modified and embellished, and in turn leading to a gentle, lyrical *coda*.

The second movement is a *Scherzo,* pure and simple, replete with *Trio* (somewhat more tranquil), designed to show off not only the contrabassoon's agility but its sense of humor. Toward the end—like the broom of the *Sorcerer's Apprentice*—the contrabassoon bifurcates, figuratively at least, into *two* contrabassoons, the soloist being joined by a colleague in the orchestra, and the movement comes to a merry "double-your-pleasure" ending.

The third movement, *Lento,* alternates various lyrical passages with long melodic lines in the contrabassoon, all under a constant string *tremolo* pedal-point. The aforementioned *cadenza* leads directly to the *Finale, allegro vivace,* in which the contrabassoon's material is varied and embellished in two successive recapitulations, producing in form a rather brief *Rondo*. The last of these variations elides unexpectedly into a quiet *coda,* which turns out to be the ending of the first movement, gently reorchestrated and ornamented.

In addition to the solo contrabassoon, the score of the *Concerto* calls for 4 flutes (2 doubling piccolo), 4 oboes, (2 doubling English horn), 3 clarinets (1 doubling E-flat clarinet, 1 doubling bass clarinet), contrabass clarinet, 3 trumpets, 4 trombones, tuba; timpani (four players); celesta (doubling piano); harp; and strings.

In Praise of Winds
(1981)

I chose the form of the "symphony" for this work because I believe, in contradistinction to much of what one was told in recent decades by leading composers and tastemakers about the obsolescence of many nineteenth-century forms, that the four-movement symphony is still a viable form and vehicle, far from exhausted. This is actually my third symphony, No. One being my early *Symphony for Brass and Percussion* (1949-50). My Second is a work entitled *Symphony* (1965) for full orchestra; and *In Praise of Winds* is No. Three, with No. Four on the way, a four movement Symphony for Organ—perhaps the first of its kind since Sowerby's *Organ Symphony* of 1936.

It is not that the symphony and sonata form types were *inherently* obsolete and dead; it is rather that much of the music the "professional avant-gardists" of the 1950s and 60s were writing and promoting did not fit these forms. The conclusion reached was that the forms were at odds with the new styles of language, when actually the reverse was true. It was much of the new music, coming out of Darmstadt and Warsaw and the Cageian philosophy, that was still-born; but it was difficult to convince anyone of that in those days.

In any event, the symphonic form seemed to me appropriate for a work of major size (both in instrumentation and duration) and one worthy of the occasion of its premiere: the celebration of the 100th Anniversary of the School of Music at the University of Michigan, and secondly the twenty-first National Conference of the College Band Directors National Association.

The first movement, with its somewhat somber and portentous opening, soon develops into a bright allegro. The opening thematic material experiences a number of transformations, while at the same time light-textured passages, exploiting the many different choirs and timbres of the large ensemble, alternate with climaxes in which its full aggregate force and power can be heard.

The second movement is mostly slow and serene, exploiting quartal harmonies and the more pastel colors of the ensemble. There are also substantial solos for trumpet and horn, long-line expressive passages of the kind twentieth century music has mostly eschewed (at least since Mahler's late works).

A virtuoso Scherzo follows, in which three shiny-bright high register chords unleash—three times—the breathless running Scherzo proper. A milder Trio section intervenes before the final re-orchestrated run of the Scherzo material, thus yielding an AABA form.

Movement four is cast in a Rondo form, in which both the main thematic material (fanfare in character) and the interspersed "episodes" are constantly varied, either in orchestration or in substance. Indeed, the third and fourth appearances of the Rondo "theme" are treated as a waltz and an up-tempo jazz section respectively. The opening flourishes of the movement are used ultimately to expand into a broad and stately coda where the basic "harmony," on which virtually the entire work is founded, is heard as an ultimate cadence and summary.

One final thought: there are still far too many composers who will either not write for the band (read: deign to write for the band) or if commissioned to do so, will regard it as peripheral to their main concerns, at best a sort of *pièce d'occasion*. Whatever the merits of *In Praise of Winds* may eventually be judged to be, my intent was certainly to write a major, substantial, "serious" work for this extraordinary medium. My fondest wish, then, is that my symphony, despite its gigantic demands, will show others what a remarkable vehicle of musical/artistic expression the modern band is and can be.

On Light Wings for *Piano Quartet* (1984)

My Piano Quartet is one of many recent works in which my primary concern is with the forging of a recognizably personal harmonic (melodic) language. For it seems to me that the rediscovery of a readily identifiable language, which can communicate all aspects of human expression, is what is most lacking in late twentieth-century music. To me it borders on the miraculous that composers such as, say, Brahms, Tchaikovsky, and Dvorak, all using the same harmonic vocabulary (i.e., the same repertory of chords, for example), could nevertheless create a *totally* personal language, whereby one can never confuse one bar, one phrase of any one of these composers with one of the others!

Whether we can rediscover such personal vocabularies—dialects, accents, if you will—in our time remains questionable. But that is our task, I believe; and whether audiences, after a seven-decade-long onslaught of a mostly amorphous impersonal atonality and innumerable detours of neo-movements of one kind or another (neo-classicism, neo-romaticism, neo-tonality, etc., etc.), would recognize and appreciate such a personal language nowadays is an even greater question.

The answer does not lie in neo-solutions. I am not interested in "returning to tonality," for example. Nor is "minimalism" a viable response to the problems of contemporary music, for it excludes too much. I *am* interested in reclaiming many of the *values* and *qualities* of the past.

I think we must find our personal identity *within* the *total* existing pitch language, acquired over some seven hundred years of Western music history, including but not limited to that of atonality—just as Brahms and Tchaikovsky worked with the fullness of the acquired language of *their* time. The answer lies not in *reduction* (as with the "minimalist" school) but in *selectivity:* the best choices out of a vast multiplicity, perhaps even infinity of options.

Thus the music in my Piano Quartet results not so much from "experimentation" or any kind of radical attempts at "originality," but rather from an urge to identify a personal language which will be recognizable (i.e., distinct from others) and also capable of communication to a relatively sophisticated audience.

The tempo markings of the Quartet's four movements—Impromptu, Scherzo, Fantasia, Bagatelle (With Swing)—give a clear indication of the different moods and characteristics intended in each. These are further modified by the over-all title *On Light Wings,* a metaphor for the work's generally light character. The last movement attempts once again to bring the strings into the world of modern jazz—an idiom to which this category of instruments (and instrumentalists) has been curiously resistant.

Shapes and Designs (1969)

Shapes and Designs is another work in what has turned out to be an on-going concern of mine with the possibilities of translating visual shapes and designs into musical structures. Earlier works along this line are my *Seven Studies on Themes of Paul Klee* and *American Triptych,* the latter being based on paintings by Jackson Pollock, Stuart Davis and Alexander Calder. This preoccupation is based on the premise that certain—not *all,* by any means—but *certain* visual or geometric shapes can be translated into musical designs, if one equates the vertical aspect of a visual design with range or register, and the horizontal visual aspect with time.

In this work, each of the four movements represents a musical realization of certain basic simple visual designs. The four movements are: I-Intersecting Triangles, II-Links, III-Arcs, IV-Wedges.

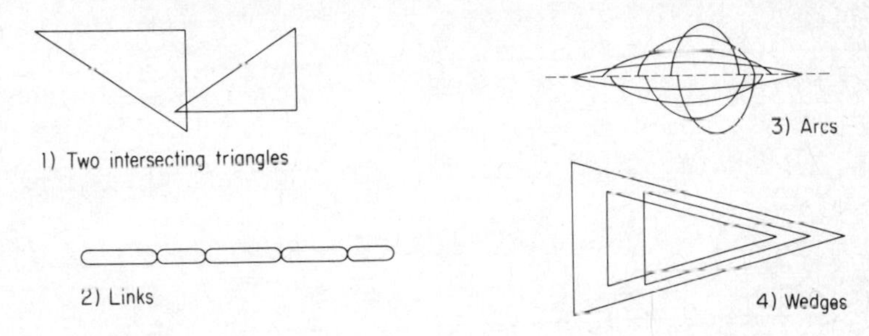

In the first movement the left triangle is represented by the strings, starting with a single pitch in the high register and gradually fanning out along the hypotenuse in a downward broadening pattern to the low register of the basses. Before the strings break off, low woodwinds enter—at the point where the two triangles intersect—and in a gradual rising shape build a second triangle. The "height" side of this triangle is represented by the abrupt cut-off of the entire orchestra.

The second movement consists of thirty-one measures of music for percussion and strings (played percussively). At the conductor's discretion, these "blocks" of sound are repeated in a variable pattern which in its entirety will form a chain made up of various-sized links of this material. Each link is delineated by the clear return to measure one.

In the third movement (only approximately represented in the graph above), a series of arcs—both upward and downward—radiate from a sort of fulcrum: middle C in the strings. The curve of each arc is determined by the size of the musical interval employed. Thus there are quarter-tone arcs (both rising and falling) in the muted horn and flute, respectively; semitone arcs, also in the flute and open horn; major seconds in the clarinet and bassoon; minor thirds in the oboe and trombone, etc. Each arc has its own rate of unfoldment, from the slowest (the quarter-tone arcs) to the fastest (major thirds) played by the violins and tuba at the apex of the piece.

The fourth movement consists of three superimposed wedges, one for each instrumental choir in the orchestra. All three wedges are identical in shape and content, except for variations in size. Thus the largest wedge, played by the strings, is mirrored by the next largest one in the brass, but being smaller, both vertically and horizontally, it also comprises a smaller range and is of shorter duration.

28

Still More Program Notes—on Others

Another selection of program notes, this time on other composers and works. The first item was written in connection with a "monster" concert (entitled 800 Years of Music*) conducted by Schuller at the New England Conservatory of Music with the Conservatory Symphony Orchestra on April 4, 1973. Beginning with a "modern" transcription of ancient Japanese "court and ceremonial music," Gagaku—a work called* Etenraku*—and ending with György Ligeti's* Lontano*, the concert spanned a millennium of musical history, in between touching upon a variety of "major stations" in the development of Western music.*

The note on Monteverdi's Orfeo *was written in connection with Schuller's revival of that opera, again at the New England Conservatory (in 1971) in what was at that time—and is to this day—one of the rare presentations of this masterwork in an authentic performance, using Monteverdi's original instrumentation and authentic period instruments.*

The performance of Purcell's Dido and Aeneas, *double-billed with Ravel's* L'Heure Espagnole, *took place in February 1974. Debussy's* Pelléas et Mélisande *was presented at Harvard's Loeb Drama Center in February 1970 by the New England Conservatory Opera Theatre on four successive evenings, sung alternatively in English and French.*

The two notes on Stravinsky's operas—the double-bill of The Nightingale *and* Oedipus Rex, *and* The Rake's Progress*—were presented in connection with performances Schuller conducted in January 1970 and May 1973 respectively.*

Eight Hundred Years of Music

Apart from (hopefully) giving listening pleasure, this concert has an educative purpose, in that it presents an overview—highly selective, to be sure—of the extraordinary breadth and longevity of our musical heritage. If I have reached for our first work beyond the realm of European music, it is to emphasize that even the extravagant riches of our own European musical traditions provide only a limited view of the *total* world of musical cultures, both ancient and contemporary.

Etenraku—meaning "music brought from heaven"—is one of the oldest examples of the so-called Gagaku music, the "court and ceremonial music" of the ancient Japanese royal court. This particular form of music is believed to have been brought from China in the early 700s, replacing the earlier Gigaku music, buddhist ritual masked dances accompanied solely by flutes and drums. The instrumentation of the Gagaku is much richer, consisting of the main melody instrument, the *hichiriki* (a type of oboe); the *oteki* (a flute) whose function is to embellish in descant fashion the main melody; the *Shô* (a kind of bamboo mouth organ); the *koto* and *gakubiwa,* plucked lute- or zither-like instruments; and as the percussion instruments *Taiko* (a huge suspended bass drum), *shoko* (a gong), and *kakko* (a cylindrical drum).

The music is basically pentatonic, although this harmonic scaffold is enriched by chromatic alterations and by copious use of glissandi and bent notes. A most remarkable feature of *Etenraku* (and most Gagaku music) is its—for our Western ears—extraordinarily slow and stately tempo, maintained sometimes for hours. Its meter corresponds to our 4/4, which is clearly delineated by the *kakko* with its unique "accelerando" rhythm every alternate measure. Further rhythmic punctuations occur in the bass drum and gong. The music also divides into episodes of eight-bar phrases, and in essence consists of a kind of variation form, in which the main melody is slightly varied and altered in successive episodes. The flutes, one and two octaves above the oboes, play in a near-unison with the latter, occasionally veering off the main tune and providing strangely unexpected dissonances.

The instrumentation of *Etenraku* for modern orchestra was made in 1931 by the Japanese composer and conductor Hidemaro Konoye. It is a faithful translation of the original Gagaku orchestra sound. The hichiriki (played in multiples in Gagaku) is transcribed for five to six oboes, plus further doubling in E-flat clarinet, trumpets, soprano saxophone, and violas. The oteki part is given to the flutes and piccolo. The shô, providing a six-part harmonic accompaniment of high "spheric" sounds, is given to six violins playing without vibrato. Harp, piano, bassoons, horns, and pizzicato cellos reproduce the original *biwa* and *koto* accompaniment.

From the "heavenly music of the spheres" of ancient Japan, it is an abrupt leap to the medieval austerities of fourteenth-century France and the music of Philippe de Vitry. In fact, the immediate juxtaposition of eighth-century Japan and fourteenth-century Europe offers us a startling comparison between the already highly developed, harmonically rich, rhythmically complex "orchestral" music of the Orient and the rather "primitive" three- and four-part organum of the early Renaissance in Europe.

When we think of fourteenth-century music, we think automatically of Guillaume de Machaut. But between Perotin (thirteenth century) and Machaut, there appeared another great composer and poet whose music is as yet little known, but whose revolutionary influence on the course of European music can be considered at least as equal to Machaut's: Philippe de Vitry (1291-1361). Certainly he was considered Machaut's peer by Petrarch who called Vitry "poeta nun unicus Galliarum." Vitry's neglect is attributable to the fact that none of his compositions were known until 1921, when a number of works in the Codex of Ivrea (Northern Italy) were identified as his. This discovery enabled musicologists in turn to establish certain motets in the *Roman de Fauvel,* a fourteenth-century collection of poetry and music, as being Vitry's. Since then, further research has established Vitry as not only an important forerunner of Machaut, but one of the major theoreticians of the *trecento,* and indeed the inventor of the term *ars nova.*

It is perhaps not too extravagant to think of Vitry as the Beethoven of his time. He was certainly an epoch-making innovator, whose contributions to musical history retained their influence for many centuries. First, he created the motet form of the *ars nova.* Apart from the purely musical-technical considerations which distinguish Vitry's motets from the earlier *ars antiqua,* they also represent a music which is independent of religious doctrine and court control. Vitry's music no longer serves only ceremonial or social functions, but is declared (by him) autonomous—art for art's sake. Indeed, the church summarily forbade the use of Vitry's motets in religious functions of any kind. (Ironically, twenty-five years later, Vitry was made a bishop of the diocese of Meaux.)

Secondly, Vitry established the notion that the composer should no longer remain anonymous. Before his time, musicians "composed" in virtual anonymity, adhering to strictly controlled musical forms and limitations imposed by the church. Their music, if "published," was not identified with their name, and it was only accidental when a name like Franco da Cologna or Perotin became known in wider geographical circles. Vitry declared the individuality of the composer of music, just as Beethoven six hundred years later pushed artistic freedom and individualism even further.

These "declarations of independence" have corollaries in Vitry's music itself. Prior to his time, there existed a limited number of motet "types," which were more or less pre-determined in respect to motivic and rhythmic content. Vitry's motets, on the other hand—even the earliest ones from 1316—are not only new in many respects, but each of his motets has its own personality and musical characteristics. Thus, Vitry was the first European composer to attempt *personal expression,* this expressiveness being at the service of and closely linked to the sense of the text. Vitry's texts, too, are no longer merely the scriptures, but highly rhetorical, polemical, often quite complex dissertations on sacred and even secular subjects. In short, Vitry's music embodies the very spirit of individualism which motivated the Renaissance.

Fourth, Vitry was the creator of the iso-rhythmic motet: a structural principle by which rhythmic patterns are strictly maintained while the actual notes (pitches) are permitted to vary from period to period.

Both motets to be performed tonight are iso-rhythmic with the vocal parts retaining their rhythmic patterns (lively and flowing), and the instrumental parts other patterns (in longer durations, more akin to *cantus firmus*).

Despite the intervallic and "harmonic" limitations, both motets are quite different and individual. In *Vos qui admiramini,* we hear not only unusually "relaxed" lilting triplet rhythms, but passages verging on "diatonic harmony." *Lugentium*—a work dedicated to Pope Clemens VI in 1342, when the papacy was in residence in Avignon, France—is unusual for its motivic imitation (a technique totally unknown at that time) and its use of a free *Tenor* part.

Another leap of three hundred years brings us to Claudio Monteverdi (1567-1643), and to the familiar world of harmony and chromaticism.

That Monteverdi is one of the great masters and innovators of our musical heritage need hardly be emphasized here. Suffice it to place the three works performed tonight in the context of Monteverdi's development as a composer of madrigals and other small vocal forms.

The earliest, *La piaga,* is from the Fourth Book of Madrigals (1603). Thus it predates *Orfeo.* It is a remarkable excursion in harmonic invention, expressing the anguish of the heart in pungent chromatic progressions, spicy dissonances, contrasting moods, and almost Wagnerian lyricism. The use of harpsichord and bass is authenticated by practices of the time, and in fact, Monteverdi who was himself an excellent player of the bass viol, frequently doubled as singer and instrumentalist.

Con che soavità is from the Seventh Book of Madrigals (1619) and is scored by Monteverdi for "solo voice and nine instruments with continuo accompaniment." This, then, is no longer a madrigal, but one of the many works which Monteverdi composed (along with Peri and Caccini) which signaled the demise of the madrigal form and foreshadowed the solo cantata.

Il lamento della ninfa—another mini cantata—dates from 1638 and comes from the collection of "Madrigals of War and Love." Here again, the solo voice is predominant, accompanied by vocal trios and *basso continuo.* The middle section "lament," in the form of a passacaglia, is one of Monteverdi's very greatest inspirations, with its natural grace, inexhaustible melodic invention, and utter simplicity. These features contrast markedly with the brief Prologue and Epilogue, laden with chromatic "dissonances" and even one purely "atonal" passage.

Beethoven's Pastorale Symphony is, of course, one of the monuments of the great Master's art. Arguments about Beethoven's "programmatic" intentions have raged for over a century, with the disputants usually quoting only those of Beethoven's recorded remarks which will substantiate their particular bias. I believe—to state my own—that neither a strictly pictorial, descriptive interpretation *nor* an "impressionistic" approach fit all phases and aspects of this symphony. Perhaps the genius of this exquisite work is that it eludes such precise confinement and definition. Certainly the first two movements concentrate on the *feelings* evoked upon "arrival in the country" and at a "brookside." Even the bird calls at the end of the second movement are more evocation than imitation. The "peasant merrymaking" is one step closer to reality, particularly in its sardonic comment on country musicians (the stubbornly uninventive bassoon, trying in vain to make the only three notes he knows fit to the oboe's shepherd tune). Then in the Storm Scene, Beethoven is no longer describing *feelings,* but in fact, a storm—raindrops, thunderbolts, and lightning strokes and all.

As Eugen Jochum pointed out to us at a recent rehearsal, a storm has no personality, no human expression; it is a physical event, and Beethoven makes that distinction by treating the storm almost as an abstraction. In the final movement, we return to the world of human beings, and the work ends by striking a remarkable balance between expression of feelings and pictorialism.

Among the many "miracles" of the Sixth, I would cite its expanded use of the winds in the orchestra, almost pushing the strings into an accompanimental background, its thematic cohesiveness from movement to movement (note how many of the themes and motives feature the descending fourth); and beyond all that, its uninhibited luxuriating in the mood and capturing the moment. In this respect, it stands unique among the Beethoven symphonies.

Erwartung (Expectation) stems from undoubtedly the richest period of productivity in Schoenberg's entire life. Created in white-heat inspiration in the astonishing time of seventeen days, *Erwartung* came only a year after Schoenberg had broken through the bounds of tonality in a startling series of works initiated by the Opus 11 *Piano Pieces*. It may forever remain a mystery how a composer who was in effect exploring—singlehandedly—a whole new tonal territory, forging a whole new language, could create a work of such incredible inventiveness as *Erwartung* with such absolute authority and such a sure sense of the inevitable. One possible explanation is that the very freedom which Schoenberg experienced in 1909 after shaking off the bonds of tonality—a freedom of choices and decisions unprecedented in the entire history of music—this glimpse of a vast, new, open future is undoubtedly what most urgently inspired him. But that explains the drive, the energy that motivated him. It does not explain the extraordinary skill and creative imagination with which *Erwartung* is set forth.

On purely statistical terms, the amount of notational detail, the melodic/harmonic/orchestrational, mosaic-like fragmentation, the fact that the score contains no literal repetition, the fact that in twenty-seven minutes of music there are 171(!) tempo changes—all this requires a prodigious control of one's *métier,* and is staggering enough. But that does not take into consideration the beauty of *Erwartung*'s melodies, the incomparable richness of its harmonies, the daring, luminescent use of the orchestra, and the extraordinary sense of timing, of concentration of form and structure.

Schoenberg called *Erwartung* a "monodrama in one act"—a monodrama for one soprano and orchestra. It must surely be the first Freudian music drama, featuring an "interior monologue" in the Joycean manner. In turn, it inspired numerous other operas, most particularly Berg's *Wozzeck,* which—as even Berg was happy to admit—is *Erwartung*'s offspring.

If we look for antecedents to *Erwartung,* we search in vain. Surely nothing in operatic or symphonic literature quite prepares us for the radical perfection of this score. At most, one can catch glimpses of some of the more daringly "neurotic" passages in Strauss's *Salomé* and *Elektra* in Schoenberg's first opera. And far in the background of memory, there lingers the impact of Wagner's *Tristan* (particularly the hallucinatory scenes of the Third Act); and in a way—as Robert Craft, I think, once put it—the woman in *Erwartung* is Isolde, but fifty years later—"with a nervous breakdown."

The opera, when staged, is in four scenes. Essentially the "plot" concerns a woman in search of her lover. She is in a feverish emotional state, on the verge of hysteria, given to hallucinations, seeing apparitions in the forest. Her utterances are almost always in a broken, fragmented manner (to which the music surely forms a close parallel). Her lover has failed to visit her for many nights, and she has now gone in search of him. In her anguished state, at one point, she stumbles against a tree trunk which she mistakes for a body (Freudian symbolism?). Eventually she approaches a house, dark and shuttered. But her foot suddenly strikes a man's corpse. It is her lover. In a dialogue with the corpse (shades of *Salomé*), it is revealed that he had taken another woman. In her final moments of dementia, the woman wanders aimlessly off with the phrase "Ich suchte" ("I was seeking"), as the orchestra also vanishes into nightmarish nothingness, in what is surely the most remarkable musical ending ever conceived by a composer.

György Ligeti's mastery of contemporary musical concepts and techniques has been established in a number of extraordinary works, including *Atmosphères, Aventures,* and *Nouvelles Aventures,* the *Woodwind Quintet,* the recent *Melodien,* and our work of tonight, *Lontano.* The latter's progenitor is Schoenberg's *Five Pieces for Orchestra,* Opus 16 (1908), the first truly a-thematic piece, built solely upon harmonic and timbral considerations. Likewise, *Lontano* is a mosaic of static sounds, formed out of as many as eighty different instrumental parts (each string player having individual parts to perform). It is a work in which the subtle play of colors and the almost imperceptible shift of single pitches to clusters or other "harmonic" aggregates constitute the totality of the "action." That all of this is organized in internal patterns—almost geometric in visual design when notated—incorporating "canonic" entrances on single pitches, dynamic canons, and similar acoustical configurations, may be meaningless information to the listener uninitiated in the complexities of today's musical composition. But as in all great works of art, the quality of Ligeti's "construction" is balanced by a keen car for the subtlest of sonorities and a superb sense of timing as to how long a static situation can be maintained without boredom setting in. It is this balance of a keen mind, a refined ear, and superb taste that places Ligeti far above the hundreds of "sound-texture" composers of the last decade.

Orfeo

He who would perform Monteverdi's *Orfeo* faces, before he even begins his first rehearsal, a veritable barrage of obstacles and problems, enough to discourage even the most ardent of devotees. For it is a fact that as yet there exists no definitive critical edition of the work, and one cannot simply go into a store and buy an "authentic" edition of the work, because none really exists. The composition of *Orfeo* falls into that early pre-Bach period when innumerable aspects of the work were not precisely defined or written down by the composer. There was, for example, no single set of instruments for which *Orfeo* was composed, and insofar as certain instruments are designated at all by Monteverdi, he does so only occasionally in the score. In scene after scene the choice of instruments is simply left to the performer or to the limitations—human and instrumental—of a given performance situation. Moreover, questions of style and the most vexing problem of all, that of vocal and instrumental embellishment or ornamentation, are left open to personal interpretation by the performer. The original "score," printed in 1609—and reprinted in 1615—is a mere skeleton of the work and even figured bass realizations are so rare in the score as to be the exception rather than the rule.

To compound the problem for today's performer of *Orfeo,* most of the instruments for which Monteverdi wrote the score are no longer in use and cannot, for example, be found in the modern symphony orchestra. One must therefore have access to cornetti, sackbuts, baroque trumpets, gambas, lutes, positive organs and many other instruments that are available today only in large metropolitan music centers, and even then mostly in contemporary recreations of those instruments.

We are fortunate to have at the New England Conservatory, under Daniel Pinkham's direction, a Collegium Musicum which specializes in the performance of early music, whose members own or have access to most of the instruments called for by Monteverdi. Thus this particular problem could be rather easily overcome in this instance, and we are pleased to present these *Orfeo* performances as played by an "orchestra" which comes very close to the original "authentic" *Orfeo* orchestra. It consists of:

6 recorders	2 violas	1 lute
1 bass racket	1 violoncello	1 regal
2 cornetti	2 contrabasses	2 positive organs
5 baroque trumpets	2 harpsichords	1 harp
5 trombones	2 virginals	3 bass viols or viole da gamba
4 violins	1 bandora	

Monteverdi calls for three arch lutes or arch citterns. Since there is only one such instrument on the entire eastern seaboard of the United States, and since that instrument (a contemporary reproduction) was not available to us, we are presenting *Orfeo* with one lute and a *bandora,* a close relative to Monteverdi's *chittarone.*

Having our "authentic" Monteverdi orchestra assembled, we now tackle the problem of acquiring the music—a score and a set of parts. Whereas in 99 out of 100 cases you merely ask a publisher to sell or rent you a set of orchestral materials, in the case of Monteverdi's *Orfeo* you discover—if you don't already know it—that there are no parts available that could in any sense be called "authentic."

Since 1881, when the first "modern" edition of *Orfeo* was published, there have appeared some 20 to 30 other renderings of the opera, most of which are contemporary transcriptions for the modern 20th century symphony orchestra. Some of these, like Ottorino Respighi's, are extraordinarily effective on their own terms, but they are, of course, of no use to anyone who wishes to perform *Orfeo* with cornetti, lutes, viols and such. Many other editions make various kinds of compromises in the interest of being "practical," substituting oboes for cornetti or trumpets for alto trombones. Still other editions attempt to be pseudo-scientifically authentic and, in the dogmatic concentration on musicological authoritativeness, end up being unusable in performance. Indeed, often they are not intended to be performed but merely to be studied and to reside in someone's archives. Yet another group of editors think they can improve upon Monteverdi by re-barring most of the opera and realizing its continuo parts in the style of a Bruckner or some other latter-day composer.

One soon realizes that in one way or another even those editions that attempt to be "authentic" are unsatisfactory, and one is left with but one alternative: to make one's own edition.

This I have done, basing all decisions on a close reading of the two editions published in Monteverdi's own lifetime, and filling in otherwise with judgments which attempt to strike a balance among several imperatives: (a) to adhere not only to the letter but also to the spirit of the score insofar as this can be gleaned from the *Urtext;* (b) to approach the work with all due respect but somehow to also find a contemporary equivalent in performance practices—since in any case we cannot be absolutely sure of precisely how the work was intended to sound; (c) to render the work in a style consistent with Monteverdi's other early works and to allow *within* that style considerable room for improvisatory and ornamentational freedom; (d) in short, to perform the opera not as a museum piece nor as a relic of the past, but as a timeless masterpiece: a work of music theater which is as contemporary today as it was in 1607.

Monteverdi's *Orfeo* score consists of many brief segments of music alternating between purely instrumental pieces (Sinfonias, Ritornelli, Dances, etc.), Choruses, innumerable Recitatives, and one or two pieces which can even be described as Arias. The melodic instruments (violins, recorders, cornetti, etc.) are confined to ensemble or choral pieces, while the whole range of continuo instruments is deployed in the Recitatives. Monteverdi makes a further division of his orchestra: the strings, recorders, harpsichords, organ and lutes are associated with the "real" world, the world of shepherds and nymphs in which Orfeo resides (Acts I, II and V), while the cornetti, trombones and regal are confined to the underworld scenes of Acts III and IV.

With this as a broad outline to follow, I determined the instrumentation of each segment on a combination of two premises: (a) to keep the instrumentation—as per Monteverdi's own often-stated principals—as varied and as contrasted as possible; (b) to maintain as much as possible specific associations between certain characters and certain accompanying instruments, almost in the sense of a *leitmotiv;* (c) and in general to exploit the astounding richness of sonorities intrinsic to Monteverdi's orchestra. How often, after all, can one hear two harpsichords, two lutes, a harp, organ and two gambas all playing a continuo part together?

I began my "edition" of *Orfeo* with the arduous task of checking the two *Urtext* editions of 1607 and 1615. As might be expected, while the errors in the former edition are corrected in the latter, entirely new mistakes crept in. Very often a given passage appears in different versions, both technically correct and possible, but making a clear-cut choice difficult. This problem arises time and time again in the continuo parts, where Monteverdi simply does not indicate whether a given chord is to be in the minor or major, or whether it is to be played as a fundamental triad or a first inversion. Such decisions are bound to be very often a matter of taste and choice, but this factor of variance in harmonic interpretation is inherent in Monteverdi's concept of the opera. presumably the performers and musical director were expected to be skilled in such matters. Improvisation was the life and soul of performance in the 16th and 17th centuries, and performers were expected to realize the harmonies and figuration extemporaneously, with the sole aid of the bass and melody lines. To a large extent we have incorporated this technique in our performances. Generally speaking I have followed two principles in regard to the continuo parts: (1) to keep the improvisations basically simple and uncluttered while allowing for special harmonic effects when the libretto or the mood of a scene calls for them; (2) to give the continuo players considerable latitude in situations where a single instrument is accompanying a singer, but confining them more in sections where *several* continuo instruments play simultaneously. It must be remembered that in Monteverdi's day the art of improvising and ornamenting was developed to a very high level, whereas today very few performers are skilled in this style of improvisation. We are therefore obliged to reacquire these skills through diligent practice and rehearsing. Sugges tions from the musical director tend to ensure a more cohesive approach to the realizations.

Monteverdi uses the quintet of baroque trumpets only in the opening Toccata, which is a fanfare-like piece undoubtedly designed by the composer to call the court at Mantua to order and to herald the beginning of the operatic entertainment. The Prologue follows, in which *La Musica,* accompanied by harp and organ, sings of the power of music to calm troubled hearts, to inflame minds, to inspire the soul. A *Ritornello* is interspersed among the five stanzas *La Musica* sings. This Ritornello I have orchestrated so as to present, in different combinations, all the instruments of the orchestra, passing in review, as it were, while *La Musica* sings of Music itself.

In Act I, Monteverdi sets the scene of the opera: the world of shepherds and nymphs who have gathered together coming from the mountains ("Lasciate i monti") to hear Orfeo, the most famous singer in the land of Thrace, and to celebrate his marriage to Euridice. It is a happy music enlivened by spirited dances, gentle Ritornelli and noble choral hymns to the Sun and to Hymen, the god of marriage.

Act II begins with an extraordinary sequence of dances, shepherd's songs, instrumental Interludes with violin and recorder duets, and finally a Dance and Air sung by Orfeo which is remarkable for its alternating 6/8-3/4 meters. The gay festivities are interrupted by *Messagera* (the messenger), who tells Orfeo of the death of Euridice. Monteverdi's setting of this scene is startling in its bold harmonic juxtapositions. In a plaintive Recitative ("Tu se' morta") Orfeo vows to go to Hades and bring back Euridice. The shepherds and nymphs comment in a chorus full of bitter dissonant harmonies, "Ahi, caso acerbo" (Ah, bitter fate). In a rare instance of thematic correlation, the bass line of this Chorus is an exact transposition of a solo vocal line sung earlier by one of the shepherds. Monteverdi dexterously converts the erstwhile melodic line into a fundament line, which acquires new and pungent harmonies in the process of transformation. Later in that Chorus, Monteverdi indulges in a marvelous bit of musical imagery. The words "che tosto fugge" (which soon disappears) are sung in fast, i.e. fleeting eighth notes. Next the words "a gran salita" (to a grand ascent) are set in rising block-like harmonies. This Chorus closes with a little syncopated *falling* figure to the words "il precipizio e presso" (the precipice is near). Moments later a Sinfonia for strings and continuo instruments expresses in its daring and unexpected harmonies all the bitter sorrow of this scene. The Act closes with the Ritornello from the Prologue, a final recollection of Orfeo's former happy world prior to his descent to Hades to retrieve his beloved Euridice.

Cornetti, trombones and regal introduce Act III in a splendid Sinfonia. We are not only on the border of the underworld but also in a different world of sonorities: the darker harsher sonorities of "brass" instruments. In the celebrated Aria "Possente Spirto" (Oh, powerful spirit), Orfeo invokes the power of music and song to persuade Caronte to let him cross the River Styx. Here again Monteverdi presents the instruments from the orchestra as soloists to aid and abet Orfeo's pleas. In succession and interspersed between Orfeo's highly embellished aria (including Monteverdi's own embellishments), we hear two violins, two cornetti, a harp in an elaborate solo, and finally a string trio. With a rather humorous touch in the midst of Orfeo's bemoaning his fate, Caronte goes to sleep on him, signified by the composer in a stately chorale-like Sinfonia, previously heard with five trombones but now played almost like a lullaby by dulcet strings.

Act IV is remarkable for its Plutone sections, their stark harmonies set for four trombones and regal in our version. Later, in the famous scene in which Orfeo leads Euridice away from Hades, Monteverdi uses all his harmonic skills and instrumental know-how to dramatize the situation, alternating in a highly fragmented sequence various combinations of accompanying instruments to delinerate each step in Orfeo's downfall.

The Prologue-Ritornello leads us to the shepherds' world at the beginning of Act V. Orfeo returns to Thrace without Euridice, now turning to the mountains and to nature for consolation. An Echo which occasionally answers Orfeo's plaintive song attempts to console him in his anguish. Apollo, Orfeo's father, then appears and chides his son for his bitter complaints and his giving-in to his emotions. In a florid duet Apollo and Orfeo ascend to Heaven singing the praises of virtue, eternal joy and peace. A sprightly Ritornello and a Chorus which comments on Orfeo's now-happy existence follow; and the operatic pageant ends with a Moresca, a wild uninhibited dance.

Those of us who have worked on *Orfeo* these many weeks and months are left with an overwhelming admiration for the remarkable simplicity and economy of Monteverdi's language. A very limited repertory of harmonies, occasionally stretched to its outer limits, serves to express a vast range of emotions and moods. Somehow one is compelled to wonder if we today, in our complex society and with our complex musical techniques and concepts, have not lost something, perhaps—like Euridice—never to be regained. It is a sobering thought. And in a way it is what Monteverdi's *Orfeo* is all about: the power and celebration of music and of sound. No nobler testament to music has ever been given.

Dido and Aeneas

There is much that is astonishing and mysterious about Henry Purcell. Obscurity hides his ancestry, birth and indeed, even his parentage. He was a natural genius who died at age 36 (when Bach was four years old); spent, as far as we know, all of his life at Westminster Abbey as composer, organist and singer (earlier in life as organ tuner and part-time copyist); created profusely in a limited number of areas, i.e., church anthems, songs, fantasias and sonatas for strings, a vast storehouse of incidental theatre music (for some 40 odd plays), a long list of official choral pieces (odes, welcome and birthday songs, wedding pieces, and such), a half-a-dozen semi-operas, and one opera: *Dido and Aeneas,* thought by many to be his greatest single achievement.

And in fact, *Dido* was a turning point in Purcell's life. After years of setting to music unbelievably dreary texts, licentious poems and other "poetic" drivel, Purcell found in *Dido* his true calling and spent the next (and last) six years of his life writing music for the theatre. But even in *Dido* Purcell was haunted by an impoverished plot and libretto. Yet the prodigality of his musical inventiveness was obviously unperturbed by such lack of textual substance, and indeed Purcell had had a great deal of practice during his professional life, lavishing his extraordinary talents on texts much worse than Nahum Tate's elliptical libretto.

Purcell is at his greatest, ironically, when he is working with verbal imagery. His music transcends the texts. His poignant and pungent harmonies—more original and daring than Handel's, for example—serve entirely to heighten poetic and emotional states. His use and mixing of minor and major tonalities, matched only by Monteverdi, seem at first indiscriminately arbitrary—and would probably rate poor marks with orthodox harmony teachers—but are in the end so telling in underscoring the opera's moments of emotional stress. Unorthodox, too, are his phrase lengths. *Dido and Aeneas* is full of five-bar phrases which work, although they feel vaguely uncomfortable at first to the unaware performer. Purcell often saves these "irregularities" for moments of greatest dramatic tension, as, for example, in the opera's crowning achievement, Dido's lament "When I am laid in earth." Here the ground bass consists of five bars (repeated eleven times), while Dido's vocal line comprises two nine-bar phrases, then another of eleven bars which is, when reiterated, foreshortened to ten.

The tension created by imposing the irregular upon the regular is one of Purcell's favorite devices. The inexorably "walking" bass lines not only give the music a marvelous flow and momentum, but their very regularity allows the composer to set up, as it were, the subtle shifts of meter in the melodic lines, the unpredictable harmonic contrasts, in short, the astonishing freedom within a very strict framework, like the *chaconne* or *ground.*

And still, all is in proportion. Each character is etched sharply and succinctly. Each scene, each word is calculated in moments. There is no padding, even taking into account the many purely dance numbers which are ingeniously integrated into the plot and music. We are fortunate to have the dance elements of *Dido* represented in what we believe to be accurate reconstructions of seventeenth-century dance notations. We are grateful to Dr. Julia Sutton, Robert Fenwick and his dancers for supplying this touch of authenticity.

L'Heure Espagnole

Though Maurice Ravel was raised a true Parisian, his Basque ancestry and his birth in Ciboure in the southwestern corner of France, a few kilometers from the Spanish border, express themselves unequivocally in much of his music. Most overtly, we hear this in a number of relatively early works, *Alborada del Gracioso* (1905), *Rhapsodie Espagnole* and tonight's opera *L'Heure Espagnole* (both 1907). (Some might think to include *Bolero,* but its Hispanic qualities are rather heavily filtered through American jazz and the Parisian "music hall.")

L'Heure Espagnole was Ravel's only opera as against a number of ballets and the opera-ballet *L'Enfant et les sortileges*. It is possibly an oversimplification, but, I think, nevertheless containing a kernel of truth, that Ravel could not turn easily to opera because he was not a man given to wearing his heart on his sleeve. And one suspects that is why his only operatic foray is a short comic opera, in fact, a farce. It plumbs no great emotional depths; its characters are just this side of caricatures: silly, self-centered, pompous, shallow buffoons. But Ravel's etching of them is cool and full of malicious humor and wit. And everything runs as precisely as the clocks in Torquemada's shop.

At a time when the world had not quite recovered from the Wagnerian revolution and Strauss was bombarding the senses with the hysterias of *Salome* and *Elektra*—it is staggering to think that Schoenberg's *Erwartung* came just two years later (1909)—it is perhaps no surprise that Ravel's *Spanish Hour* was not an immediate success. Only after its revival in 1921 in Brussels was it more cordially received; without, however, ever becoming a standard operatic staple. But then in the diamond horseshoe world of opera, emotionality is preferred to sensibility.

And *L'Heure Espagnole* certainly has a great deal of the latter. We hear it in the highly personal style of Ravel's vocal declamation. Every verbal nuance has its musical counterpart, at once subtle and piquant. The orchestration, masterly as always with Ravel, does not indulge in mere instrumental effects, but with spectacular virtuosity and wry wit, cartoons the characters for us. Here again, this is done subtly, virtually unnoticeably: the orchestration never intrudes. And though the orchestra is large, it is used sparingly.

One suspects that Ravel had the most fun with the introduction, a delightful evocation of a small-town clockmaker's shop. Ravel loved mechanical toys, music boxes and the like, and in the opening of *L'Heure Espagnole* he could really indulge his imagination. Listen for the clicking of the clocks (three metronomes set at different speeds), the different pitches and patterns of the clock chimes, the various mechanical figures (marionettes in the celesta and harp; the toy bugle in the stopped French horn, the rooster played on the reed of the contrabassoon, and, of course, a chirping bird in the piccolo). It is all quite magical—a lovely, childlike fantasy, a wisp of a vision that is gone almost before we can appreciate it.

Withal, Ravel is first and foremost a classicist, not a sensualist, and not even an impressionist. There is an uncanny, almost scholastic neatness in *L'Heure Espagnole,* and it brings to mind an apt remark once made by Ravel's friend Vuilllermoz as to the distinctions between Debussy and Ravel: "There are several ways of playing Debussy's music; but there is only one way of playing Ravel's."

À Propos Pelléas

"On ne dit pas, on suggère." Mallarmé's famous words, espoused by Debussy and propagated to the stature of a dictum by Pierre Boulez, undoubtedly found musical fulfillment in *Pelléas et Mélisande.* But these words and the sense they contain have also been—too often—detrimentally applied to Debussy's *chef d'oeuvre* by those who would emasculate it and desensualize it. The traditional approach to *Pelléas* has been to envelop it in vaporous clouds of "suggestion," to anesthetize its drama and music with layers of cotton soaked in the balm of impressionism and symbolism, to drain its content by under-statement or non-statement, to destroy its form and harmonic lucidity by willful "interpretation." In presenting *Pelléas,* not only in the context of public performance but also in the context of an educational effort on behalf of our Conservatory students, we have striven to return to Debussy's original conception and intentions. We have tried to strip off—

in so far as they were foreign to the work—the accumulated encrustations of tradition and have returned to a concept of the work in which both the frame of reference of Maeterlinck's drama and Debussy's classic sense of form, proportion and structure are reasserted. In these efforts we were fortunate to have constant access to Debussy's original *Pelléas et Mélisande* manuscript, of which the New England Conservatory library is the proud possessor. I might mention parenthetically that numerous notational errors, discrepancies, wrong notes and rhythms—many of them undoubtedly dating from the hectic pre-première rehearsal period in 1902 and still not weeded out of the presently available scores, parts and piano reductions—were corrected in the process of this reassessment.

The return to Debussy's "original" has been considered not only out of a sense of operatic one-upmanship or even a musicologist's obsession with the Urtext, but simply to rectify a long standing misconception regarding much of Debussy's mature work. For the wonder of *Pelléas* is that it transports the symbol and the dream into the realm of absolute music. Debussy, the perfect anti-Wagnerite, returns the supercharged world of Wagnerian symbolism, of Wagnerian chromaticism, of Wagnerian declamation, to the concise, precise, clearly delineated world of French classicism. The methods are no less classic. The recitative prosody of the French classic opera of Rameau is revived, and Rousseau's famous demand that vocal lines move in small intervals, avoid the unnecessary exaggerations and affectations of the aria, is respected and adhered to. Yet the *Pelléas* language is modern, and traditional harmonic functions are virtually suspended. Wagner's leitmotif-principle is partially accepted, but also transformed: the leitmotif is no longer used—as in Debussy's own brilliant quip about Wagner's tetralogy—like a photograph identifying each character, but rather as thematic material to be constantly varied, transformed, and redefined.

The extraordinary discretion and subtlety of Debussy's musical language in *Pelléas* have led many unwary performers to distort both its form and content. What is often mistakenly considered prissy and fussy in Debussy's score, is in reality a unique sensitivity to sonority, timbre, and texture, and a sensual delight in exposing his musical materials in an ever-changing kaleidoscope of colors, shading, and nuances. Thus, what appears at first glance to be fragmentary and discontinuous, is, in reality, merely a mosaic-like conception of form and continuity.

It is, therefore, the performers' obligation in *Pelléas* to connect what appears to be *dis*connected and fragmented, to collect the thousands of musical details, the hundreds of minute forms and structures, the dozens of relatively longer formal entities, into one total five-act form, in which the myriad implied and explicit relationships—particularly of tempo—are preserved and delineated.

Debussy does not make it easy for us, for he uses no metronome markings in *Pelléas;* and the hundreds of *retenus, plus animés,* and *rubatos* provide the conductor with a formidable task, similar to the one a captain faces in trying to keep a ship on an even course in very heavy seas. There is a great temptation for singers in this opera, because it contains no arias and virtually no long-lined vocal melodies in the traditional operatic sense, to elongate and blow up the numerous unaccompanied recitatives into little *ariettas,* thereby impeding the natural flow of the music and distorting Debussy's very precise sense of form.

Like all great masters, Debussy hears the different sections, movements, or scenes of his works—and *Pelléas* is no exception—not as unrelated and separate segments, but as integrated parts of a large entity. Tempos and movement—as in Mozart or Beethoven symphonies—are internally related, emanating from a common rhythmic source, in the sense that different tempos are merely variants of some all-embracing base tempo, related by a common denominator—what we today call in composer's terminology "metric modulation."

But form and continuity are not only determined by or expressed in rhythmic terms. Harmony—in the larger structural sense—can tell us as much about musical form as meter and rhythm. And, since Debussy replaced the traditional functions of diatonic harmony by a more flexible and open-ended type of tonal organization, the harmonies of *Pelléas* are, in fact, our main clues to tempo and continuity. In fact, harmony and tempo constantly fluctuate *in phase* with each other. For example, in a music in which there are, except for the five act-endings and a handful of other places, otherwise no final harmonic resolutions, where harmonies are, in other words, in a constant state of unresolved flux, tempos must flow in analagous or related fluctuations, and yet, in the end, collate into large formal entities, like scenes or acts.

The need for maintaining flow and movement in *Pelléas* is, moreover, substantiated by Debussy himself when he writes that he "wants the action never to stand still, on the contrary to move without interruption. It is for this reason that I tried to free myself from the parasitic and traditional operatic formulas." Accordingly, we have returned—at least in our English version of *Pelléas*—to Debussy's "original" and, for its time, very daring conception of presenting Maeterlinck's drama in thirteen scenes with virtually no interruptions between scenes. Debussy was correct in assuming that the unbroken momentum of sharply contrasting scenes, following each other quickly, almost abruptly, would heighten the dramatic impact of the opera. But at the première in 1902 the scene changes could not be technically managed, and Debussy was persuaded to compose a number of orchestral interludes, which, though at times extraordinarily beautiful, do impede the dramatic action.

In general, Mr. Strasfogel and I have tried to establish a parallel between the music and the staging, not only in regard to questions of tempo and pacing, but in relating a more clearly delineated rendering of *Pelléas*'s musical structures (formal, harmonic, etc.) to equally sharply defined dramatic characterizations. Above all, there is the constant reminder—even by the stage set itself—that this human tragedy is placed in a dramatic landscape of endless lonely forests, bleak countrysides, and starving peasants, in an atmosphere permeated by death, and a world in which people desperately try to relate and communicate with each other, a world more real than perhaps ever conceived in Maeterlinck's "mythical" setting. Debussy himself was quite clear on this point: "*Pelléas et Mélisande* is a drama which, despite its dream-like atmosphere, is much more human and real than many of the so-called real-life plays"—by implication the *verismo* operas of his time.

How wonderfully Debussy states both context and content in the first six bars of music: the stark, somber harmonies of the eternal forests, uninvolved and noncommittal; followed by the warmer, vibrant chromatic harmonics of Golaud's theme—the former static and uninvolved, the latter moving and engaged.

Here in these six bars Debussy presents the music and drama of *Pelléas* in germinal form. Not only is essential thematic material presented, but the two stylistic-harmonic poles of his musical language are set in relief: austere antique medieval harmonies contrast with the sensual, refined chromaticism of his own invention. And, above all, it is as if the first four bars are a symbolic representation of the frame, i.e., the physical setting, while the next two bars represent the picture itself, the *dramatis personae*. It is my hope that Mr. Strasfogel and I will have succeeded in bringing out the constant interplay between these antipodal elements.

Le Rossignol *and* Oedipus Rex

A double bill of Stravinsky "operas" may well cause a raised eyebrow or a wrinkled forehead with inveterate operagoers. For, in truth, neither *Le Rossignol* nor *Oedipus Rex* is an opera in the traditional or orthodox sense. The former consists of a series of fairy-tale *tableaux*, while the latter is as much an oratorio as it is an opera. Indeed, the composer and his collaborator, Jean Cocteau, called *Oedipus* an "opera-oratorio" or a "scenic oratorio." It is significant that the work was premiered in 1927 in *concert* performance.

But if our interest in these works is renewed today, it is partly because the much-belabored crisis—some say demise—of opera has forced us to turn our attention again to something called "music theatre." And certainly *Le Rossignol* and *Oedipus Rex* were early and significant explorations—albeit totally different and created nearly two decades apart—toward a new music theatre, which might rivivify opera and at the same time bring that venerable genre into a closer relationship with the new language of music as it unfolded so unexpectedly in the teens and twenties of this century. These two "operas," precisely *because* they lie somewhat outside the "pure" traditions of grand opera, not only provide the stage director of today with superb opportunities to use new means of stagecraft but—because of their open-endedness, operatically speaking—also permit a free translation from "opera" into truly contemporary "music theatre." This, of course, brings these pieces well within the realm of training and education in music—the New England Conservatory's primary business—particularly as these works represent relatively recent historical attempts in grappling with the problems of contemporary opera, and thus may provide us with insights and clues as to what direction the music theatre may take in the *future*. In the sense that these works give the young student performers of the Conservatory a living experience with two examples of the recent "operatic" *past,* as reinvested with attitudes and techniques of the *present,* they constitute, I believe, a most fitting educational enterprise for a forward-looking school opera department. They, in fact, teach our young performers certain lessons about "opera" which more traditional operas could *not* teach. As public performance we hope, of course, that these operas—the one a brilliant *spectacle* (in the French sense), the other an ageless legend and ritual with obvious modern parallels—will entertain, uplift, as well as educate.

Consider the diverse and unusual requirements of these two "operas," so totally different from each other that the uninitiated listener would hardly guess they come from the same creative hand. Both works involve substantial chorus participation: in *Le Rossignol* in a traditional manner, very Russian, very beholden to Moussorgsky's *Boris;* in *Oedipus* in a completely new (in 1927) and particularly Stravinskyan manner. In both works there is a minimum of dramatic action—certainly in any theatrically naturalistic sense. *Le Rossignol* is all magic theatre, while *Oedipus* is more ritual than drama, more moving sculpture than stage action. *Le Rossignol* is highly fragmented—virtually devoid of the traditional operatic forms, arias and narratives—and ultimately the main protagonists are not the singers but the orchestra, which performs a good two-thirds of the music without benefit of voices. There are no *big* parts for the singers on the scale of a Donna Anna or a Salomé or a Cavaradossi. It is an ensemble opera essentially.

Oedipus does have its "arias," of course, but they evoke much more the spirit of Handel and Bach oratorios than that of nineteenth century grand opera. And of course the chorus plays a central role, at least equal to that of the main characters of Sophocles' drama.

Both operas eschew traditional operatic acting. *Le Rossignol* involves much mime and dance movement in its non-verbal sections, while *Oedipus*—with its concept of archaic stylizations—certainly calls for a very special kind of "singing actor," requiring the controlled movements of a dancer. Both works can thus be viewed as early and important forerunners of the present-day interest in a music theatre in which actors and mimes play a primary role, and in which the pioneer work of Brecht, Merce Cunningham, Alvin Nikolais and Grotowski have played an influential role.

Stylistically both operas again offer great and quite diverse challenges to the performers. In *Le Rossignol* there is the problem of assimilating the various influences on the young Stravinsky: Debussy, Scriabin, Rimsky-Korsakov in the firt act; Moussorgsky, Schoenberg and, of course, his own *Rite of Spring* in the second and third acts. Perhaps one should not speak of a Schoenberg "influence," but in any case the nightingale's music in the second act is very close to the free atonality of Schoenberg's *Pierrot Lunaire,* the premiere of which Stravinsky heard in 1912 in Berlin. The well-known stylistic split between the first act of *Le Rossignol,* written in 1908 and 1909—that is, *before* the *Firebird,* the commission of which, by Diaghilev, interrupted the composition of *Le Rossignol*—and the other two acts, written in 1913 and 1914, after the *Rite of Spring,* is in my view not as much a problem as certain Stravinsky biographers have made of it. It is not just a *post facto* rationalization to say—as Stravinsky has in his *Chroniques de ma Vie*—that since the "action" of *Le Rossignol* does not really start until the second act, it would not be entirely illogical if the music of the prologue-like first act were somewhat different in character. The forest with the nightingale, the fisherman's simple song, the whole refined atmosphere and poetry of the opening of Andersen's fairy-tale could not be dealt with in the same manner as the Chinese Emperor's court, with its bizarre and exotic population, its tintinabulation of bells and bright sounds, its hustle and bustle.

Much more problematic in my view is the internal fragmentation within acts, which reaches its peak in the latter part of the second act—with the aforementioned Schoenbergian *Chanson du Rossignol,* the Japanese Envoy's music, the music of the mechanical nightingale, the Emperor's banishment of the real nightingale, and so on. Stravinsky himself was aware of this problem even as he composed *Le Rossignol.* He felt he was not ready yet to write an opera, and that ballets were much more his rightful medium. Oddly enough, *Le Sacre* also consists on the surface of an accumulation of brief episodes and terse structures, many of them with neither beginnings nor endings (in the proper or traditional sense), and marked by extraordinarily abrupt interruptions and changes of course. Yet somehow in *Le Sacre* an inner thread binds these many formal links into a grand totality, overwhelmingly perfect in design, while in *Le Rossignol* essentially the same approach does not work nearly as well. It is therefore the conductor's and performer's obligation to bring an over-all continuity to this fragmentation, and in this lies perhaps the greatest performance challenge of *Le Rossignol*.

The stylistic challenges of *Oedipus Rex* are of an entirely different sort. I speak not of the general problems—primarily for the stage director—of dealing with the monumental, the mythic, the purposely archaic and "static" qualities of the work. As conductor and therefore as "teacher" of the student orchestra and chorus, I was particularly concerned with finding a sonority, a sound, a texture, in short a style that is peculiar to this period of Stravinsky and which probably reaches its apex in *Oedipus Rex*. I refer to the fact that on the one hand the music obviously rejects the "romantic" attitudes and sounds of the nineteenth century. On the other hand, as much as Stravinsky espoused the "new objectivity" and "neo-classic" postures in vogue in the 1920s, the romantic heritage was not—all protestations to the contrary—entirely shaken off, nor indeed so easily eradicable. It peeks through the many harmonically chromatic passages, through the instrumental writing (especially in the strings), indeed through the entire fabric and texture of the piece. Though Bach and Handel are his essential models, they are heard through ears that had experienced the nineteenth century rediscovery and romanticization of Bach, particularly in its French strain. That is why the many subtle touches of Verdi in *Oedipus* do not seem out of place.

We used to think that Stravinsky's music from this period was to be played in a dry, *sec*, perhaps even expressionless style—where only the facade of the music was important, a facade presumably devoid of content. Indeed not so many decades ago the Schoenbergians still preached such notions about Stravinsky in a pretty contentious manner. Today, I believe, we know better than that, if for no other reason than that already the *Symphony of Psalms* (1930)—with its uniquely rich and stately harmonies, its lyric polyphony, its inner glow and fervor—shows us that Stravinsky's music was not to be confined within such a narrow stylistic conception. *The Rake's Progress* (1952)—with which *Oedipus* now reveals in retrospect a markable affinity in style—would be the other striking example to disprove the myth of Stravinsky's music as being all craft and no soul.

Clearly the dry, humorless, *sec* styles of the early performances of neo-classic Stravinsky was not the answer. (I might add as a footnote that those of us old enough to have been musically weaned on the old

78 recordings of Stravinsky, mostly recorded in the twenties and thirties in Paris with French musicians, were all misled into believing that they represented the true Stravinsky style and sound. But those recordings were the result of accidents of time and place, and as such point up the inherent prejudicial danger in recordings.) We know today that the performance tasks in Stravinsky's music are much more complex, not reducible to such absurd limitations. We can play this music neither with the colorless sound we once thought appropriate nor—to be sure—with the sound of Brahms or Wieniawski. What we must find is something between these two extremes, a distillation of the sound that combines pure sonority with an *inner* glow, an *inner* expression. The lyricism of *Oedipus* must come from within—distilled out of the harmonies, the melodic lines, the rhythmic propulsion of the ostinato figures. *Oedipus* cannot be played with a store-bought sound or, for example, with a vibrato imposed from without, overlaid onto the tone, as it were. It must rather come from a precise understanding of and feeling for such things as harmonic and intervallic relationships. There is very little room for "interpretation" in the nineteenth century understanding of this term. Stravinsky's harmonic language—which in *Oedipus* is essentially tonal, yet with its own very contemporary flexibility and linear independence—*provides its own comment* on the musical heritage from which it derives. It *interprets* Bach or Verdi or Mozart for us; it *is* its own interpretation and requires no further.

To walk this sonoric-stylistic tightrope is, of course, not easy. But I believe it is one of the disciplines young performers must learn and one which is frequently ignored, even at professional performance levels.

In some ways the chorus writing in *Oedipus* presents the most unusual musical problems. First of all there is a curious ambivalence regarding the function of the chorus. It is, of course, essentially the concept of the Greek chorus, as commentator and representing the people of Thebes, which Cocteau and Stravinsky sought to revive. But beyond that dramatic role, the chorus functions alternatively (1) as a primary *vocal* voice which the orchestra accompanies, (2) at other times *instrumentally* as part of the orchestral texture, and (3) occasionally even in a secondary role to the orchestra. The first problem, then, is to find a balance and continuity between these constantly shifting functions of the chorus in the over-all musical fabric.

Secondly, in respect to the part-writing for the chorus, an analogously shifting situation prevails. Stravinsky's choruses in *Oedipus* alternate continuously between four-part, three-part, two-part and unison writing. This creates very special balance problems. Since the number of singers remains constant, the number of voices per part varies greatly depending on whether Stravinsky divides the whole chorus into two, three or four parts or writes for unison chorus. These shifts in balance and density occur abruptly, almost arbitrarily, and may appear in all four forms even in the span of one or two bars. When one realizes that over and above these internal balance problems there is the over-all problem of the shifting role of the chorus in the *total* musical texture, one can perhaps appreciate the magnitude of the task.

I find that most of the time the choral writing in *Oedipus* performs harmonic functions, rather than melodic. That is, it fills out and supplements the harmonies which are sometimes incompletely sketched in the orchestra. This is an unusual approach, for most oratorio or opera choral writing is, of course, harmonically or polyphonically self-sufficient. The orchestra may double these musical ideas or it may play its own independent parallel role. But rarely are the two used in an intertwined manner where neither orchestra nor chorus is wholly independent and where, in addition, the degree of intertwining may vary greatly. For example, we frequently find in *Oedipus* that a particular vocal part, let us say the second basses (in a four-part division), is the only carrier of an important pitch in a particular chord or harmony. This vocal part has to be made to blend with the instruments and the other three voice parts so as to create a balanced harmony, while not destroying the linear functions of any given line. Conversely, an instrumental part, let us say the English horn, is the only carrier of a note which crucially completes or defines the chord, the other notes of which are represented and doubled in *both* the chorus and the orchestra.

Perhaps the most experimental aspect of our approach to *Oedipus* is the use of some slight electronic amplification and modification of the chorus sound. I wanted to delineate very clearly three distinct timbral units: the orchestra, the solo voices, and the chorus. This is to disassociate the chorus, the populace of Thebes, bystanders and commentators, from the more individual and personally engaged solo protagonists. It is made possible by the fact that but for one or two small exceptions, the solo voices and chorus do not sing simultaneously. It also made possible the multiplication and amplification of the choral voices into a more monumental massive force.

One could not often encounter two operas that are so different and yet the creation of the same composer. Though the instrumentation is virtually the same in both operas, the sounds the two orchestras produce could not be more disparate. Brilliant and virtuosic, sophisticated, *raffiné* in *Le Rossignol*, the orchestral sonority of *Oedipus* is by contrast classic, sparse, economical, often stark and dry, and in its big moments monolithic. It is perhaps worth noting that (by coincidence) *Oedipus* was the first Stravinsky work to use the full orchestra since *Le Rossignol*. How his orchestral palette had changed in the intervening twelve years! He now was applying for the first time to the full orchestra the lessons learned in *L'Histoire du Soldat*, the *Symphonies for Wind Instruments*, *Pulcinella* and other works, instrumentationally determined partly by war economies and partly by a radical change in composers' attitudes regarding the orchestra—attitudes which had discarded the gigantic ensembles of Mahler, Strauss, Schoenberg's *Gurrelieder* and, of course, of Stravinsky's own *Rite of Spring*. *Oedipus* is, in many places—particularly in its handling of the woodwinds—much closer to Bach's *St. Matthew Passion* than to his own earlier works like *Le Rossignol* or *Le Sacre*.

Our work in the past months has been guided by the concerns and challenges expressed in these notes. We hope that these efforts will have served to focus and clarify the important contribution both works have made to the lyric stage.

The Rake's Progress

I must confess at the outset that Stravinsky's *Rake* is one of my half-dozen favorite operas. Having played the horn in its premiere at the Met, participated in the first recording (with Stravinsky conducting), and subsequently conducted the work at the San Francisco Opera, I find that the *Rake* has truly become a "repertory opera" for me—one that is not just an "esoteric" piece occasionally revived to spice up a season, but a masterpiece in the great lineage of operatic durables.

I suppose it was inevitable that a work drawing upon the talents of two of the twentieth century's greatest creative spirits (Auden and Stravinsky), leaning in turn on one of the keenest social observers and artists of the eighteenth century (Hogarth), would make its mark in operatic history. But I don't think that anyone in the early 1950s suspected that Stravinsky would write something much more than a superbly crafted score in the manner perhaps of the two symphonies ("in C" and "in three movements") or the delightful *Scènes de Ballet*. What Stravinsky gave us instead is not only an overwhelmingly moving musical account of the Rake's path to perdition (and finally salvation) but a work whose profundity is rivaled only by the *Symphony of Psalms* (if we stay within the so-called neo-classic period). But even beyond these two achievements, *The Rake's Progress* is the sublime culmination of Stravinsky's "neo-classic" involvement: it constitutes the pinnacle of achievement for that musical philosophy and language, and it is the greatest and last paean to "classical" harmony in the twentieth century.

It is this latter aspect which is probably most rarely realized. I suppose this is because most performers think of Stravinsky—even to this day—as the great rhythmic innovator and revolutionizer of our time. And under the aegis of rhythmic clarity—also because rhythmic problems stubbornly remain the single most difficult performance obstacle—Stravinsky the harmonist is constantly short-changed, subverted, indeed often *per*verted. His mastery of extended tonality in the *Rake* goes far beyond his expected skill and craftsmanship. It is uncanny, almost unanalyzable in its profound inventiveness.

All this happened, of course, at a time when Stravinsky was already beginning to eye serialism seriously. Indeed, the *Rake* turned out to be Stravinsky's swan-song farewell to tonality. Again it seems to me that this was inevitable: the perfection of the *Rake*'s harmonic/melodic language could not be surpassed in that particular style. There really seemed no place further to go. And no need; Stravinsky had said all that could be said in that particular language.

What more touching and more beautiful tribute to spring, to nature and to love do we know in opera than the opening A-major duet between Anne and Tom? This gentle "innocent" music, with its ingeniously subtle modulations, seems to me to set the whole tone of the opera. And how masterfully the music then traces each stage in Tom's ne'er-do-well path, finally coming full circle with the poignantly simple music of the asylum! It is the music of the opening duet—distended, clarified, distilled, apotheosized—delineating not only the Rake's progress but the progress from reality to *ir*reality.

In between, we have the ribald brothel episodes; the brilliant Act I closing *Cabaletta* aria; the profound loneliness of the trumpet solo depicting Anne's arrival in London; the bizarre music of Baba the Turk—pompous, awkward, and temperamental; the grim graveyard scene with its somber string quartet introduction, its ingenious use of the harpsichord; and finally the terrifying power of Nick Shadow's blasphemous curse.

It has often been noted—and freely admitted by Stravinsky—that the spirit of Mozart, particularly *Cosi fan Tutte,* hovers over the *Rake's Progress.* This is indeed so. However, to overemphasize this factor, as is often done, is to do injustice to Stravinsky. The *Rake* is not a mere neo-classic return to the stylistic features of *Figaro, Don Giovanni,* and *Cosi.* Reincarnation would be a more apt tribute. There is nothing *pseudo* about this music, nothing manneristic, and there is not one note that isn't immediately discernible as pure Stravinsky. What remains of Mozart is so thoroughly subsumed into the very essence of Stravinsky's language that we hear it as something new and original. It is Mozart revisited, to be sure, but by a genius who has shown us aspects of Mozart we have never seen before.

I have spoken of the *Rake*'s extraordinary love affair with harmony. Listen for this, for it pervades every nook and cranny of the score. It is just possible that these ravishingly beautiful sounds, filtered and savored by one of the greatest musical ears of all time, were not only the most appropriate way with which to set Hogarth's bitter-sweet tale, but—more profoundly—Stravinsky's own bitter-sweet farewell to tonality.

29

Concerning my Opera The Visitation

This article, originally written in German and entitled "Worum Es Mir In The Visitation *Ging," was commissioned by the German music journal* Musik im Unterricht" *(Music in Education) for its October 1968 issue. Translation into English by Mr. Schuller.*

AN OPERA IS usually the product of many musical as well as extra-musical impulses and interests. I can only articulate here some of the more special concerns that motivated the writing of *The Visitation.* Without the impetus of Rolf Liebermann's commission "for a *jazz opera"* (for the Hamburg State Opera), the work would probably never have been created. For Liebermann's idea struck a particularly resonant chord in me, since I have spent a good part of my musical life in jazz (as well as in classical music and opera—the latter as first horn for fifteen years at the Metropolitan Opera). It was also perhaps inevitable that my numerous attempts in previous instrumental works to fuse contemporary jazz and classical concepts would eventually lead to a similar stylistic fusion in a "dramatic" work. Once that idea had germinated in me, it gradually began to take shape from rather specific and (for me) inevitable textual, musical, and philosophical points of departure.

I knew from the outset that the use of jazz, in whatever form, could only occur in connection with a subject matter, a text, a libretto relating to the Negro, whether as a people or as a symbol, or as a specific individual in a specific historical or dramatic situation. This was not merely a matter of finding an appropriate literary or dramatic subject, but rather a matter of honor: a view which is perhaps difficult to appreciate in Germany, where jazz is not an indigenous music, and whose opera public rarely encounters jazz as a serious, living music.

To comprehend this point, at least philosophically, it is necessary to remove the many misconceptions about jazz that still persist, especially among "classical" music lovers, even as to *what jazz is*.

In relation to my opera it may be easier to indicate what jazz is *not*. It is in this instance not necessarily a "popular" music, nor is it definable as any of the many commercial derivatives of jazz, which in Germany are, along with jazz, officially labelled as "light music," i.e., by implication non- or un-serious music. Jazz in my opera is a modern, partially improvised, spontaneously creative art music, measured by the same kind of rigorous, demanding disciplines and aesthetic criteria as any other art music. To put it simply: I was not interested in grafting jazz into an otherwise "normal" contemporary opera style, much less to "jazz up" my score with a few deft "jazz numbers." I *was* interested in a real integration of fundamental contemporary jazz and classical concepts, techniques and idioms, in the same sense that the Third Stream concept had already some years earlier postulated such a fusion.

Jazz as an art form—in all of its various stylistic manifestations—is the creation of the American Negro. Its entire development from a simple folk music to a national entertainment and dance music, and thence to a sophisticated art music was initiated and directed along its creative path by the Negro. In some profound and unequivocal way I felt that my opera must pay tribute to this fact, implicitly or explicitly. But once that basic tenet was in place, a variety of choices and options still remained to be decided upon: would the opera be based on a literary or dramatic work by a Negro author; would it be about the Negro race or an aspect of its history; should it deal with a specific individual fate; or might it interpret the Negro in America as "symbol" of all oppressed minorities; or, finally, should it perhaps be an opera about the many "tragic figures" that populate the history of jazz itself?

After a three-year search and after many bitter disappointments (in connection with copyright restrictions, translation rights, etc.), I decided to "transpose" Kafka's novel *The Trial* to the American realm and particularly the American race problem. From the outset the idea was not to write a Kafka-opera as such, but to articulate the Kafkaesque elements in the American "Negro problem." Perhaps some people—especially Middle Europeans, well versed in the Kafka literature—may see in this transformation a simplification and narrowing of Kafka's original conception. This may even be to some extent a justifiable argument. But I believe that something is also gained by this musico-literary "transposition" onto a different social terrain. For it made possible the focusing of Kafka's novel in a specifically "dramatic" form: as it were, to translate the unreal and eerie in Kafka's *The Trial* into specific, actual, everyday experience. If properly staged, the opera's identification with a more specific subject could gain in intensely felt realism what it may lose in symbolistic power.

This seems to me critically important, because rather few people are able to make the jump from Kafka's uncanny symbolism—sometimes so subtly handled—to everyday reality. Many readers, for example, treat Kafka's work as if it were some unreal, remote, intangible, unapproachable literary monument, not realizing that such experiences as detailed therein can also intrude upon their own lives—for which the Hitler era in Germany, the McCarthy period, or a century of racial bigotry in general in America bear abundant witness.

The "transposition" of Kafka's nightmare world to a specifically American arena is also viable since opera in no way *excludes* the symbolic and multi-dimensional. Although we know, for example, that one of the themes in Verdi's *Don Carlos* is the suppression of the Flemish people by the Spanish, it does not prevent us—partly because music is the interpretively freest and most all-embracing of artistic expressions—from seeing in these Flemish *all* oppressed and subjugated peoples. In a similar way my opera, in *its* specifics, can refer back to Kafka's more generalized situations and, by extension, to the plight of the individual in an authoritarian or demagogic society.

Further to the basic material of *The Visitation:* the protagonist Carter Jones—far from being seen as a stereotypical Negro cliché—is a complex figure, perhaps unique to the "melting pot" America. Carter Jones is not only a Negro or *the* Negro—symbol of his race—but at another level also the representation of a certain *type* of Negro, partially assimilated into white society: namely, the Negro whose racial and personal characteristics have become undefined over years (or generations) of attempts to integrate into the dominant (white) society. Carter Jones is pulled in opposite directions but, trying to remain loyal to both, is torn apart between them. He is well on his way toward the status of the "civilized intellectual" as opposed to the folksy, unbridled primitive "child of nature." He is well-educated, reads books, is involved with the arts, is seriously concerned with raising his standard of living—presumably aspiring to meet prevailing white models. At the same time he is the somewhat naïve "liberal" who, while he clings to abstract aesthetic beliefs and theories, abhors racial and social stereotyping and sloganeering. He tries to steer a precarious survival course among society's prejudices, contradictions, and complexities, hoping at least to manage a decent life in a basically indecent and hostile world. It is this societal jungle which Carter Jones in the end cannot endure, for on the one hand he has lost the combative "street-fighter" survival instincts of the ghetto Negro, and on the other hand has not yet fully integrated into and understood the ways of his white environment. He is caught midway between two worlds, and is not a true—or accepted—member of either.

As for the *music* of *The Visitation,* I am little interested in trying to explain or validate it on theoretical/technical grounds. I leave that sort of thing to the "professional avant-gardists" all around us, who love to justify their work by pre-performance analyses and intellectual hype. Least of all do I wish to sloganize the work as a "jazz opera," let alone *the* "jazz opera." It is *an* attempt—my attempt—to fuse jazz elements with a contemporary (in my case) atonal, dodecaphonic language. Indeed *The Visitation* is much less a "jazz opera" than a "Third Stream opera." The true jazz opera has not yet been written, for it will have to contain elements that have not yet been explored, above all the use in vocal roles of real, true jazz singers and an orchestra of musicians capable of improvised creative jazz. *The Visitation* is in these respects still a hybrid "Third Stream" work, employing basically conventional operatic forces (opera singers, a traditional opera orchestra, etc.) into which have been introduced a jazz ensemble and jazz-stylistic elements.

As for the fusion of jazz and symphonic traditions, I have explored this genre in a great variety of ways in many previous works. *The Visitation* is simply another attempt at a balanced symbiosis of the two traditions. But those who come to my opera with expectations of hearing a lot of "straight-ahead," mainstream jazz, will be sorely disappointed. Jazz functions in my opera in multiple ways: improvised and written (notated); twelve-tonal (or freely atonal) and tonal; leading/primary and accompanimental/obligato; soloistically predominant and psychologically underscoring. Parts of the opera have no jazz at all; others are entirely limited to it. At other times jazz is blended with the symphony orchestra and voices in such a way that the originally separate elements are no longer distinctly discernible; they have been blended into a new, unprecedented idiomatic musical alloy.

To appreciate and accept these new fusions, the listener/viewer must try to discard all previous preconceptions—on both sides—and must be prepared to hear the heretofore segregated idioms from a new perspective, in a new unity. Ears that insist on clinging to the standard stereotypes of either "jazz" or "classical music" will find little satisfaction in *The Visitation*. This applies to the typical jazz fan, unaccustomed to the sound of an opera orchestra and opera singers, or conversely to the traditional opera lover, for whom the "swing" and "sound" and unpredictable spontaneity of jazz are a remote alien world.

The alleged dichotomy between twelve-tone and jazz doesn't exist for me; it never has—certainly not qualitatively. For at least fifteen years there have been many attempts to enrich jazz improvisation by a more fully chromatic language, whether twelve-tone or freely atonal. A number of leading jazz musicians have, in their own way, used methods and techniques analogous or parallel to contemporary "classical" concepts. If one has the right jazz musicians—seven are required in my opera—the bridge between modern jazz and an essentially twelve-tone language as used in *The Visitation* is easily crossed.

However, my opera also incorporates style elements which belong to older jazz traditions, for example the New Orleans-style funeral procession of the final scene, or the rock-and-roll piece and "symphonic jazz" treatment of a pop song, both jukebox numbers in Act I, scene 3. Beyond that I have incorporated in the opera a certain jazz-historical perspective, which in turn arises out of specific scenes and dramaturgic considerations, for example the interpolation of Bessie Smith's recording of "Nobody Knows You When You're Down and Out" in the Prologue of the opera. Bessie's performance is in itself one of the early artistic monuments of jazz, but here it serves additionally to bring into focus the sociological and musical background of the opera and the interaction between the Negro's past, symbolized in Bessie's recording, and his present. As used in the opera's Prologue, Bessie's recording calls forth a series of concise visions—not only of Carter Jones's and the Negro's social milieu, but a mini-history of jazz from its African origins through the quasi-musical improvised sermons of a Baptist preacher to the Bessie Smith performance, representing the first flowering in the 1920s of jazz (and blues) as a new and distinct art form. From this historic point on, my opera takes jazz forward through not only the three above-mentioned stylistic prototypes (jazz, rock, pop) but into the whole realm of modern small-group jazz. In this way the Bessie Smith recording underscores and unifies the opera psychologically, historically, musically, and dramatically.

The music of the entire opera is based on a single primary tone row (or set) and a "family" of closely related derived sets, which are in turn variously and selectively divided between the jazz-oriented scenes and such as are free of jazz elements. Beyond this, it should be pointed out that the pitch sets here used are combinatorial in structure and content—this an expansion of Schoenberg's twelve-tone "system" developed by Milton Babbitt—and thus comprise an entire network of related and derived sets.

But above all, explications aside, I want to say that it was not at all my intention in *The Visitation* to produce some consciously "radical" or "avant-garde" work, as the European opera houses have witnessed so often in recent years, but something much more difficult: a singable repertory opera. For that is still the mostly unresolved challenge of late twentieth-century opera: to provide a singable operatic repertory that does not compromise stylistically.

Music Aesthetics and Education

30

The Compleat Musician in the Complete Conservatory

The New England Conservatory of Music in Boston celebrated its Centennial in 1967. In February of that year Schuller, then President-elect of the Conservatory, delivered the following address at the 100th birthday party. In it Schuller picks up various themes, previously and elsewhere propounded, such as his aesthetic and educational ideal of the "total musician," along with that his "global view of music," a warning about the cancerous apathy among professional musicians, and a mild swipe at the "degree mania" in academia.

IT IS MY GREAT pleasure and honor to speak to you tonight about several matters which I shall assume (as a result of your presence here) are of more than passing interest to you; namely, the Conservatory's role in music education today, a tentative prognosis of its role in the future, and my own specific views related to these subjects. It is the first chance for me to meet you, and I suppose this is a good occasion for you to find out what ideas I have on music education in America today. If I do not begin my remarks with a few well-placed jokes, as is customary at dinner speeches, I do not wish you to conclude that I am a humorless clod, but rather that the subject to which I shall address myself is a serious and difficult one, and one which I do not care to approach with opening verbal feints and dodges and fancy rhetorical footwork. Nor do I intend to apologize for the seriousness of my approach, for I feel that only through a direct confrontation with the issues at hand, can we—all of us, who are friends of the Conservatory—hope to understand these issues, and deal with them intelligently.

For let us be clear about one thing: The conservatory is a very special kind of musical institution in our society, a society which moreover is undergoing far-reaching cultural and aesthetic changes. The conservatory has—or was intended to have—a very special function in the musical community. Its very name—conservatory—indicating a place where something is to be conserved, presents the educator with a major problem; namely, how to reconcile the conservatory's basically conservative, tradition-perpetuating function with its other obligation to constantly re-evaluate those traditions, lest the conservatory become merely a museum. Conservatories have always had to grapple with containing these essentially opposite attitudes, but perhaps never has it been so difficult to strike a rational balance in these respects as in the twentieth century, when more things have happened at a faster rate than ever before, and when vast social, economic, and cultural changes have precipitated an aesthetic tug of war which shows no signs of abating.

Whereas in earlier days conservatories were institutions with a well-defined function in a relatively stable cultural society, today after a half a century of (literally speaking) earth-shaking scientific advances and cultural explosions of one kind or another, we have arrived at a situation which in the mildest terms must be described as being in a state of flux. As a result the conservatory's role has become less-defined, and indeed, many would have us believe that its position has become very precarious.

Conservatories are threatened in two ways today. One is largely economic, in that it is becoming increasingly difficult, given the inflationary tendencies of our times, to be an independent operation—just as it is increasingly difficult to be or remain a small businessman. We all know what the pressures of big business on small business can be and how unequal the competition between the two is. In the overall set-up of educational institutions, compared to the giant universities, a conservatory is indeed a small business operation.

The second reason conservatories are threatened is that in the wildly fluctuating sociological and cultural patterns of our day, the conservatory's traditional function has been questioned. In the days when it was thought of as the unique supplier of talent for symphony orchestras and faculties for other conservatories, this problem did not exist. There was a certain very specific demand for which the conservatories produced a very specific supply, and in the process, certain—again—well-defined traditions were perpetuated and handed down from generation to generation.

But today, the universities compete with us conservatories in this same area on a scale that could not have been imagined fifty years ago.

Moreover, institutions like the symphony orchestra are themselves undergoing changes, and some people go so far as to say that the symphony orchestra's survival in our changing world is also a precarious one. There are those who feel that the future of the symphony orchestra depends on whether it can adjust to the demands made upon it by a radically changing musical repertoire. There are others who feel that the symphony orchestra and the opera house should ignore all that and become museum-like repositories for preserving the old traditions. This is, for example, Mr. Rudolph Bing's point of view. Another leader of the musical community, Leonard Bernstein, says flatly that symphony orchestras can no longer cope with contemporary music, since compositional styles and conceptions have become so complex as to place the orchestra in a squeeze-play between (on the one hand) the extra rehearsal time needed to produce a reasonable performance and (on the other hand) the lack of economic funds to provide those extra rehearsals. This position if carried to its ultimate consequence also leads us right back to the orchestra-as-a-museum theory. Mr. Bernstein's position, although in my view somewhat overstated and unnecessarily defeatist in attitude, *is* not without foundation in fact. To counterbalance this somewhat pessimistic view, there is a growing awareness on the part of foundations and other subsidy programs that there is a real problem here which is beyond the control or the means of the orchestras themselves.

There is a whole body of works and of conceptions in whose develment the orchestra cannot figure, because these compositions are conceived for performing media other than the symphony orchestra; for example, the vast body of chamber ensemble music or—even more removed—the various electronic media. The symphony orchestra cannot participate in the dissemination of this music. Whether we approve or not, these media are here to stay, and the orchestra (to put it in the most charitable terms) has lost its primary position in the performing hierarchy of contemporary music.

In any case, these shifts—I could offer further examples—in the sociological structure of music have had a profound effect on the role of the conservatory, although there seems to be some question as to exactly in what way. The dilemma of the conservatory is that it is to perform an essentially conservative function in a constantly changing and forward-moving frame of reference. To pinpoint this problem more specifically, we have only to imagine a hypothetical situation in which a progressive forward-looking conservatory might prepare its students in a repertoire with which the symphony orchestras are no longer involved. I have purposely cited an extreme case, in order to show what *could* happen; and I do not consider this hypothetical case inconceivable; in fact, we find ourselves close to such a situation already today. It boils down to the fundamental question: Shall the conservatory take the lead in the musical community; or does it passively supply a demand determined by performing organizations; or as a third alternative, does it defer the whole problem to the experimental centers at the universities, and let them worry about it?

I wish there were simple answers to these difficult questions. In my view, no single, all-embracing satisfying answer appears. But it is clear to me that conservatories must approach the future with a degree of flexibility unknown heretofore. They must—as the expression goes in the fight game—roll with the punches. In fact, the conservatory must if possible perform a multiplicity of functions, at the same time supplying the demand in specific training areas and in helping to create a general musical climate. The conservatory cannot sit back and rigidly hold to an image of itself which was valid years ago but which is no longer applicable today, at least not in its entirety.

Nor does this mean that we must throw out all the old tenets and replace them by totally new ones, breaking completely with older traditions. In this regard, I can best sum up my own personal philosophy by quoting Mark Twain, who said: "Retain of the past only that which you may need in the future."

My philosophy in such questions is never one of having to do with replacing, but rather with expanding or adding to. We don't throw out the old; we add to that which is still valid in the old that which we feel is equally valid in the new. And this "old" and that "new" must be constantly re-evaluated on their own merits and in relationship to each other, as they act upon each other. These are, I believe, the principles by which a society can survive in a changing world; and a conservatory is but a small organism in a larger social structure, in a larger society.

If this sounds like expansion to you, you have heard me correctly. But I do not mean expansion physically, in the sense of new buildings or in the sense of numerical expansion of Faculty and Student Body—although I do not rule these out as possibilities in the distant future. No, I am talking about a different kind of expansion—an expansion of our thinking, of our concepts of teaching, and of our methods of training young people. But expansion without deepening at the same time is of little value. No matter how much we broaden our conceptual horizon, we must never do it at the expense of deepening and sharpening those conceptions. Here, too, I can sum up my views in a bit of home-spun philosophy: The higher the tree grows, and the wider its crown spreads, the deeper its roots must be anchored, lest the tree fall over.

In other words, in my view the conservatory must be an institution where the young musician can go to expand the range and depth of his musical perception, to sharpen and focus his instrumental capacities, and to broaden his general intellectual horizon—in short to make the best, most complete kind of professional. And I should like to make this conservatory, this particular New England Conservatory, the kind of place young students will feel they *must* attend in order to achieve that kind of professionalism.

I think that there is one idea which is central to the over-all educational conception to which I am alluding here; and that is something I would call the *total musician,* the complete musician. We have very few total musicians today. We have something called total serialization in music, and we even have a breakfast food called "Total" cereal, but we don't have total musicians; and this is something we sorely need.

We have instead specialists—or worse than that: we have musicians who *think* they are specialists, who, though they are not particularly special in their chosen area, are at the same time totally oblivious of any other area. The music world seems to be divided into two camps, each camp espousing totally opposite theories. One theory holds that you must concentrate all your efforts on one specialized area, such as learning perfectly the half-dozen piano concertos with which you can make a career, or allow yourself to be indoctrinated by one particular successful style or conception—this is particularly true of composers—and then by dint of hard practice and dogged perseverance you succeed in becoming successful. The other theory encompasses a much broader educational view. It is a view in which you try to absorb as much music across the board and in depth as possible, and try to integrate all of these ideas into a single personal conception. Mind you, I'm not after a "smattering of ignorance," as Oscar Levant once put it so well, but after a form of musical enrichment up to the maximum intellectual and emotional capacity of the individual involved. A conservatory must be able to provide that quality of enrichment and the highest level of intellectual and emotional stimulus.

From these remarks you can gather that I lean toward the concept of the total musician and *away* from the specialist, the non-musician virtuoso, although I recognize that there are frequently exceptional cases whose special musical endowments require a more specialized treatment. Please do not misunderstand me. When I talk about the total musician, I am not talking about some monstrous, perfected genius, some kind of human computer. I'm talking about something very simple. I'm talking about giving the young musicians the tools by which they can live a life in music which is rich, meaningful, and rewarding—and not only monetarily rewarding—and not mere drudgery, as is so often the case. I can perhaps put it best by telling you of an experience I had when, a few weeks ago, I had occasion to be present at several full days of instrumental auditions. I heard in that time well over fifty young musicians, all of them either young professionals or graduates or postgraduates. Of that number I am sorry to report no more than perhaps 5 percent seemed to have any idea of why they were playing music, what a musical phrase meant—indeed what constituted a musical phrase—and what the expressive and intellectual range of music can really be. For 95 percent of them it was merely a matter of pushing down certain keys at certain times, moving arms or adjusting embouchures or whatever was involved in their instrument, to perform what appeared to be a purely mechanical operation. The whole sense of the joy of music, of the beauty of music, of the ability to communicate through music, was absent. If the computer ever takes over the world of music, it will not be because this or that composer wished it so and inflicted it on an unwilling public, but it will be rather because the passivity and utter boredom of the player will have reached such a point that he might as well be replaced by the computer, for at least the computer is efficient.

When I hear that kind of audition—and this was not atypical—I become very sad. More than mad, I become sad. But it also inspires me to try and do something to prevent that kind of complete emotional, intellectual disassociation and sterility. For it is not necessary that such a thing should happen. I know from personal experience that it need not be so, that even under the most trying professional circumstances, if those roots, emotional and intellectual roots, that I spoke about earlier have grown deep enough, one need never lose one's curiosity, one's love, one's identification with music and its rewards.

And those rewards are richer today, I believe, than ever before. In addition to the nineteenth-century repertoire, we have acquired in recent decades through improved research methods a pre-eighteenth-century repertory which, along with the continual additions in contemporary music, provide a total musical feast to whet the most jaded appetite. At the same time, the development of a wide variety of new communications media provides an outlet for this expanded repertoire which would have made an eighteenth-century musician envious. The field of music and its peripherally related areas provide a range of outlets and a potential source of income beyond the wildest dreams of our forefathers. But to operate efficiently and effectively in this expanded field, the musician has to be equipped properly—he has to be the total or complete musician.

I am always saddened when I meet musicians who plod along in their everyday existence, having no understanding or love of the music they play. They are tolerant, for example, of a few nineteenth-century composers (which they usually choose on the basis of how well those composers wrote for their chosen instrument), and they are usually disdainful of all contemporary music or even of earlier composers, like Bach for instance, in whose music the intellectual quotient dares to be fairly high, and are generally speaking pretty ignorant about the incredible variety, breadth and depth of musical languages. Many of them are not even very humble about their ignorance, and of course, many of them are teachers, who thus perpetuate in their pupils the same kinds of ignorance and prejudice, though in the meantime they may indeed be playing a pretty respectable oboe or cello or snare drum or whatever.

I maintain that that type of musician—he is the spiritual mentor of the young auditioners I described earlier—is not only a dangerous cancer in the music profession, but is depriving *himself* of the joys and pleasures of music. And I say that conservatories must be actively involved in preventing that kind of musician from happening. For that kind of musician is not a complete musician—if he is a musician at all; he really seems more like a musical mechanic or automaton. I maintain further that that kind of musician cannot survive in music today, or at least he cannot survive as well as the musician who brings a more sophisticated background and point of view to his profession. It reminds me of the old business axiom: There is always room at the top. But the top today signifies not just a digital dexterity, but an intellectual sophistication, an intellectual curiosity, and a depth of perception. The reason for this is simple: competition. The music field, as I need not tell anyone in this room, has become an *extremely* competitive field. Today when a major orchestra has an opening for tuba, let us say, there are forty to fifty applicants for the audition. The implication must be obvious: Play your horn well, but have something else besides; have an extra dimension which will enable you to fill that "room at the top."

But beyond the audition level, once the musician is safely ensconced in his job, he will be much the better for it if this job continues to have a meaning for him, a means of expressing himself, a means of reflecting the impact of the music upon himself, in short the exchange of ideas and feelings between the composer and the performer, between the creative and the re-creative, and these ideas transmitted then via the performer to the listener.

Very often in discussions or arguments about the validity of contemporary music with musicians (who sometimes are more rigid in their thinking than lay people), I tend to point out a very simple fact. This is a hypothetical example involving three musicians in connection with three composers. Musician "A" likes and understands only composer "X," a representative of the romantic school; musician "B" likes and understands composer "Y" in addition to "X" ("Y" is a baroque or pre-nineteenth century composer); musician "C" likes and understands "X," "Y," and "Z"—"Z" representing the much feared contemporary music. Is there any question as to who, purely statistically, has the greater enjoyment in music; and is there any question as to whose life is more intellectually enriched by music? Can there be any question that the musician who appreciates and understands the structural perfection of the "Eroica" Symphony, who savors while he is performing the hundreds of harmonic, rhythmic or orchestrational details that contribute to make that piece one of the masterpieces of our musical heritage, receives a kind of psychic income from performing music that the musician, who is unaware of such compositional relationships, can simply never enjoy.

We forget today that the musician of the sixteenth, seventeenth, and eighteenth centuries was rarely just a composer or just an instrumentalist. If he was an oboist, he was also a composer and perhaps a pianist; if he was a composer, he was also perhaps a flutist or an organist. The creative and *re*-creative aspects of music were an integral balance in such a musician's musical constitution, and the one fructified the other. In more recent times, this was an ideal embodied and revived by a composer like Paul Hindemith, himself the compleat musician, and one can see this concept still perpetuated at the Yale

School of Music, where Hindemith taught and created a music school that in many respects is more like the ideal conservatory than many conservatories.

Forgive me for belaboring the point, but it is a crucial one, and it will have a lot to do with the kind of music our children will be listening to ten or twenty years from now. How do we produce this complete musician? Why, very simply, by the complete conservatory, of course. And that, my friends, is a conservatory which manifests the same kind of breadth and depth in quality and conception that I have been speaking about this evening. It is a conservatory where the many subsidiary disciplines, whether applied or theoretical, whether vocal or instrumental, whether individual or collective, are all integrated, aware of each other, enlightened by each other. In that way the student who goes to such a conservatory will learn to understand—and more important—absorb intuitively how all these theories and methods relate to music, which—I must remind you—we call an art in our society. The student will gradually acquire a vision of that art, in which pushing down keys at a certain moment to produce a certain acoustic result will not be an end in itself or a means of merely gaining a livelihood, but will in addition be a vehicle for expressing feelings, thoughts, ideas—those of the composer he is performing, and even (if he has earned the privilege) those of himself.

There is one other problem I would like to touch upon, because it relates directly to my ideas of the complete musician and the role that the conservatory must play to create that kind of musician. We are going through a period in American education involved with an extremely exaggerated "degree-consciousness." We have made a fetish of the degree, a pedigree, a kind of automatic approval which cannot be questioned, and we are about to do the same in music with the doctorate. We evaluate people's ability too much on whether they have a degree or not, and what *kind* of degree they have. Literally thousands of teaching jobs are not available unless the applicant has a degree, regardless of his unique, intrinsic ability. Worse than that, our universities are full of very worthy teachers who are prevented from becoming, for example, full professors and receiving the better salaries that are attached to full professorships, because they do not have a doctorate, while some less gifted person who has a doctorate moves ahead into the upper echelons. *I think we must stop this madness!* We must stop it because we are indulging here in an abstracted educational process which puts the emphasis on the number of study hours completed rather than what has really been comprehended in those study hours. We push thousands upon thousands of students through a sort of assembly-line educational process, which remains a process rather than becoming a really full complete education.

Forgive me for becoming autobiographical for a moment, but I do it only to make a point. I stand before you as one of the original drop-outs. I do not have any degrees; I do not have even a high school diploma. Now, I'm not advocating this necessarily as a road to higher education, and I am aware of the fact that times have changed tremendously in the twenty-four years since I left high school. But I have the feeling I would not have been a very good music student in, for example, the rigid programs which allow for almost no electives, which some of our schools demand. What I am trying to say is that we must develop a new flexibility in our music education, in our programs, in our curriculi, to make room for the tremendous range in student and faculty types. We seem to be in the process of doing the opposite.

For the universities, being mammoth institutions with mammoth organizational problems, perhaps this factory-type of method is the only possible and inevitable result. But I believe precisely the conservatory, a small independent music school, can show a different approach and a different result, and this in some cautious way I would like to investigate and try to do. Institutions just like people must remain flexible, or else they will atrophy.

I realize very well that it is one thing to design an abstract blueprint such as the one I have just offered. It is quite another thing to infuse this abstract with life, to make it a consistently productive force. There is only one way in which that life can be instilled in such a blueprint: and the key to that is quality—quality of faculty, quality of student. And here there is no room for compromise. Idealism does not thrive on compromise, nor does quality. And to the extent that it is possible for me to achieve this quality in a humane way, I will pursue that goal. But I am not so foolish as to think that I can do this alone. I will need help, your help—you who are devoted to the concept of the conservatory, the venerable tradition it represents, and who are particularly devoted to *this* conservatory at 290 Huntington Avenue. I ask your help and support, be it spiritual, financial, ideational, or moral. I welcome it all, and I trust that we will someday all be proud of the results of this joint effort.

31

Qualitative Evaluation in the Arts

A talk delivered in 1980 at New York University as part of a Symposium held at the School of Education, Health, Nursing and Arts Professions. Schuller's discourse deals with the complex problems of objectivity and subjectivity in the arts, and the very central and rarely discussed question of the quality *of evaluation, of criticism in the arts. The audience—graduate and doctoral students (as well as some faculty members)—consisted of majors in Philosophy, Anthropology, Humanities, and Phenomenology.*

LET ME BEGIN by reminding you that I am not a phenomenologist, not even a philosopher or a professional aesthetician or anthropologist—although I suppose it would be hard to argue that a composer, which is what I primarily am, is not a little of all of these—at least at times and in varying ways and degrees. By making that disclaimer, however, I am not intending to equivocate or make prior apologies for what I am going to say, but rather to make quite clear at the outset that I will be speaking, as I must, primarily from practical musical experience. I trust that will be useful and that my participation in this ongoing discourse will be of some value to the symposium.

The subject at hand, "Qualitative Evaluation in the Arts," is, of course, a fascinating one. Indeed it is one in which the fascination it holds for us far outweighs the degree of objectivity and assertiveness with which we can speak about it—at least at this point in time. For I believe that we don't even know all the questions yet, let alone the answers. And insofar as we do have questions, once again the questions far outnumber the answers. It is also, alas, a field populated by many—in academia and the educational field and among the critical fraternities in the arts—who *claim* to know and daily assert answers without even knowing what the appropriate questions are.

It is sobering and instructive to ponder some of the papers already presented in this series of talks and to note that often the wisest of questions leads to a morass of further questions and philosophical dead ends.

This is not in any way to denigrate the thinking and searching reflected in these papers, but rather to emphasize—and to confirm right at the outset from the vantage point of a practicing creative artist and *re*-creative performer—how profoundly complex and unyielding of easy answers the subject at hand is.

I will also say now—to get it out in the open—that as an artist and as an educator and as a concerned arts citizen, in short as what you philosophers call a "professional"—this state of affairs does not disturb me terribly much. I would, *of course,* like to help bring some order into the chaos; I would, *of course,* like to bring what we do and teach and learn and evaluate into a closer, more precise relationship with what the object of our endeavors actually is, phenomenologically speaking. But I suspect—and my deepest instincts as an artist tell me—that there are things about this subject—this "phenomenon"—which it is *not* for us to know. Of course, I don't know that either; it's just a very large hunch that, as they say in mathematics about certain theorems, it is undecidable.

As an artist—and I believe I am a thinking artist, not hopefully a mindless artist—it doesn't bother me that I don't know everything about either the creative process or its progeny. I am happy to know that it works, and that in the main I can rely on it and the way it functions. The fact that there are unrevealed and incomprehensible mysteries in the creative-arts process and in our evaluation of its products does not disturb me, although it arouses my curiosity. But I don't have to know how something works in order to use it. The point, however, is that we cannot expect absolute qualitative evaluation of a product when the process entailed in making that product is not fully understood and therefore cannot be easily evaluated. There is a rather precise correlation in that formulation.

But more of that later. Let's first try to evaluate some of my experiences and resultant findings as a professional.

All right, how *does* one—or how do I—evaluate quality in the arts (and in training in the arts or in arts education)? Acknowledging that there are significant differences between various art forms, let me concentrate to some extent on my primary field—music. The prevailing wisdom, of course, is that evaluation in the arts is bound to be subjective, indeed, cannot be objective, and is entirely a matter of personal taste, personal opinion, and background. This I dispute to some extent for I believe that there is much more that we can know and use objectively than we generally care to admit. I could argue, for example, that if there were no objective elements to be woven into the process of creating a musical performance, we would have a totally arbitrary, willful, and anarchic approach to musical interpretation. What I would like to emphasize is that in questions of how to interpret a musical work, there are many more objective elements than we generally care to acknowledge. Despite the often-discussed inadequacies of our musical notation, which are real and which we composers and performers often agonize over, it is nevertheless true that there is much more that *is* given and commonly understood than not given and misunderstood. On a scale of 1 to 10, I suspect that the number 8 is an appropriate figure to describe what we objectively know and can rely upon. That's a lot, and if used intelligently it moves us a long way toward both a representative performance and ways of measuring it in a qualitative sense.

The other 20 percent of performance represents, of course, the most interesting, refined, sophisticated aspect of performance, as well as the most personal. These are precisely those elements which give a performance what we call the "human" or individual dimensions, and which our musical notation cannot define. (The same statement could be made about the field of dance, of course.) There is, thus, a fascinating and, I think, rather precise correlation between the areas which notation cannot define, and those which performances by variable, fallible, individual human beings—each a separate phenomenon, as it were—retain as they commingle, so to speak, their talents with that of the composer's. This amalgam, fused of the creative and recreative, is a strong and healthy one. It is precisely what helps to preserve a work, a composition, as living art, which is precisely the thing an electronically or synthetically produced composition fixed on tape cannot be. This does not denigrate the electronic medium, but merely points up its differences in this one respect.

Since a notated piece of music is only an incomplete and as yet unrealized blueprint of a composer's creation, performance is a necessity before it can communicate the fullness of its content. Oh, yes, I can read a score and hear it in my head, and in some limited sense that represents a *kind* of realization and communication. But it exists in only a private and individual realm, since for music to communicate fully, it must also exist in some form of acoustical reality and in a time continuum. This can only be achieved by a performing intermediary, an individual or a group of individuals.

The performance variables that a composer encounters and must learn to tolerate are precisely the price he pays for the communicative realization of his creation. And, as I say, the measurable and definable aspects of such a creation are in direct proportion to the amount of latitude a composer must leave in order to have life breathed into his offspring.

Incidentally, I firmly reject the term "interpretation." I much prefer the word "realization." I can cite too many instances in which realization of the composer's blueprint is what is needed, not interpretation, and certainly not re-interpretation.

As an artist who has worked both sides of the street, I can make such a point. But I would like to emphasize that there is a larger, rather than lesser, amount of material that we can view objectively and therefore, in turn, measure and evaluate qualitatively. In terms of creativity and performance by human beings, that is as much as we can aspire to. To go beyond that we must go to other media, other instruments, and other modes of expression—such as electronics and the computer can supply.

So, if someone asks me how I as a musician, music producer, and music consumer *qualitatively* evaluate a performance or a work, I answer that I do so on the basis of my talent as an observer and evaluator. For the degree of variability and divergence in evaluation lies much more in the observer and evaluator than it does in the work itself or its performance.

Another way of saying this is that the mysteries about these matters are many fewer than popular opinion would have us believe. Again on a scale of 1 to 10, everything between 1 and 8 is quite clear. It is between 8 and 10 that matters get complicated and variable, and subject to subjectivity and conflicting interpretations.

For example, if a note written by a composer as an F sharp is played as an F natural, that is a clearly measurable and definable failure of realization, rendition, performance, or interpretation—choose your own term. If that F sharp is played as an out of tune F sharp, it is easily subject to qualitative evaluation (in this case also negative). If, however, the vertical and the horizontal placement of that pitch are technically accurate and correct, we still have a host of other interpretive choices to evaluate. And here we are around 8 on my scale—with such matters as dynamic level, touch, attack, duration, how we sustain the note, with what degree of steadiness or non-steadiness, timbre/color, sonority, vibrato—and, by extension, how that note relates to all other notes in the work.

Say that I know that in his *Octet* Schubert used triple *p* (*ppp*) only once, and from all the internal evidence of the work itself it was used in a specific functional way; in other words, it is not arbitrary, accidental, or gratuitous but signals a very special moment in the entire work. I also know that this special dynamic occurs in the latter part of the last movement's development section and in fact signals the gradual return to the recapitulation of the main thematic material of the movement. If I hear a performance where that *ppp* is ignored—in other words, if that moment is not the quietest moment in the entire work, then I have a firm, objectively demonstrable judgment or evaluation to make of that part of the performance.

If, however, I, as a listener or critic, do not know about Schubert's one *ppp* in that long fifty-minute work and therefore cannot, through my own ignorance, evaluate whether the performance realized Schubert's intentions, than I cannot make a qualitative evaluation or judgment, nor would I be entitled to make one.

What develops from this line of reasoning is not so much what a work or a performance achieves or is perceived to achieve, but rather what knowledge the perceiver brings to his observation and evaluation. And that, of course, is a matter of education, not in only the formal sense, but in the broader deeper informal sense of intellectual growth.

Quite often observers, who have the knowledge necessary to make such an evaluation, can agree on the conceptual and analytical information. That's, of course, the Joker in the deck; for the wide divergence of evaluative opinion occurs in direct proportion to the difference in knowledge, aural sensitivity, and intellectual perception observers bring to such an evaluation. Again the variable is much more in the eye—or in this case—the ear of the beholder than in the object beheld.

This is true even if we are willing to admit that individual and collective tastes do change from century to century, from era to era, or even perhaps from generation to generation. Everything is finally relative; there are no absolutes; and it is as useless to bemoan our fate because we can't hear J. S. Bach play and therefore know how *he* played his music and wished it to be played, as it is to assume that we have developed ideas of performance and creation today which are of necessity valuable to another time or another generation.

To return to my own criteria of qualitative evaluation, I can only say they teeter on a constantly shifting and precarious balance between knowns and unknowns, with the emphasis more on the former than on the latter, I hope. It works like this: as firm and as authoritative as I may be at a given moment in a given situation in my judgments, I nevertheless know, at the very moment of judgment, that that evaluation is impermanent, certainly not absolute—"definitive" at best only for that moment. In other words, I feel I can dare to offer that evaluation only because I am aware of its potential fallibility.

This kind of thinking relates very intimately to one crucial aspect of the creative process that is very close to me—namely, composing. Composing, seen from one perspective, is the act of making an almost countless succession of decisions. I have written the 238th note of a piece and I now must decide upon the 239th note. I must make this choice out of a virtually limitless number of options—in respect to pitch, duration, timbre, touch or attack, function, and a dozen other contextual questions. What differentiates a composer from another person is his ability to make such decisions. What differentiates a great composer from a lesser one is his ability to make such choices in such a way that both *he and we* feel the choices were not only right, both intellectually and emotionally, but *inevitable*. When we say of Beethoven's *Eroica* that it is a masterpiece—a common qualitative evaluation—we are in effect saying that every decision Beethoven made in that piece—every decision—has a sense of inevitability about it; it cannot be improved, it cannot be altered; it cannot be subtracted from; it cannot be added to; it represents a whole series of ultimate choices in a *sui generis* situation.

Quite so! But what is interesting is that any composer worthy of the name, Beethoven included, has to live with the reality that as he makes that "ultimate," "inevitable" choice—if he is so fortunate to have that gift—he already knows that in all likelihood five or twenty years hence he might make an even better (or in any case different) decision for that particular set of circumstances. Part of the composer's ability to know that which makes him a composer, is his ability to make his "best" choice at a given moment of time and to commit it to paper (or if he is a jazz improviser to instantly commit it to his instrument). If on the other hand he does not have it in him to know that, then it is likely that he (a) will not come to a momentarily final decision, (b) may not ever commit himself to his presumably "best" decision, or (c), worst of all, may constantly revise his decision or his entire work to the point where he cannot finish it or write at all. Similarly, qualitative evaluation in the arts means that decisive action is required at the very moment that it is taken, but that it is in all likelihood subject to future revision and re-evaluation.

If education serves as the basis for making qualitative evaluations, it does not mean that we can find absolutes there, but rather that it is the only legitimate alternative to suspending judgment entirely. Analogous to what we say about democracy, as imperfect as it may be, it is the best we have. You see, here we are qualitatively evaluating evaluation!

Needless to say, education, with its preordained quantitative criteria of evaluation, is beset by so many problems and deficiencies—we don't even know how, for example, we learn music—that music education, meant now as a profession, hardly holds any answers to this tantalizing riddle of qualitative evaluation.

How far off the mark all of us generally are—audience, professionals, critics, teachers, administrators—is best exemplified by a couple of other phenomena we can observe in operation every day. I spoke about music's inherent need for an interpreter, a middle-man, an interlocutor who transmits the composer's vision to an audience, who sends the composer's signal to a receiver.

In the case of a symphony orchestra this signal-sending is accomplished by a conductor and eighty or so musicians. Now music is an aural art, an auditory art. We have become so corrupted by the excessive visualization of our lives in recent times—aided and abetted by television—that the primary way we are likely to evaluate a live orchestral performance today is through the "choreography" displayed by the conductor. Look around you at the next concert and see if you find *anyone* who is just listening to the music, perhaps with eyes closed or head bent. You won't. What you will see is a hall full of people appreciating and evaluating the performance primarily (and perhaps totally) in relation to the gestures and movements of the conductor. Whether these movements relate to that composition intimately or only remotely or not at all, whether they in fact produce a response, the right response or perhaps no response from the orchestra, is scarcely considered relevant and more often than not is not factored into the evaluation.

Another analogous example—an even more complex one—is exemplified by musical pieces with programs or non-generic titles. Take the case of Strauss's *Till Eulenspiegel,* and let us postulate two diametrically opposed situations. Listener A is a knowledgeable music lover and knows not only that Strauss created that tone poem on the basis of and drew his inspiration externally from the Flemish legend of Tyl Uilenspiegel, but he even knows that story, at least in its outlines. Either he has read the original or, more likely, has read program notes delineating the work's scenario. Is his enjoyment of the Strauss *Till Eulenspiegel* enhanced by such knowledge? Does this enhancement apply more to the first time he hears the work than the fifteenth time? Is his capacity to evaluate the work and performances of it aided and extended by such prior information?

Now take listener B, another hypothetical case—a little more unusual but not unthinkable—of an aurally sensitive person who hears *Till Eulenspiegel* not previously knowing the work or even its title or underlying scenerio. Does he enjoy the work potentially or inherently less than the informed person? Must he know the story of Till's escapades in order to like, love, appreciate, understand the work? How will such knowledge or lack of knowledge affect his qualitative evaluation?

Well, we don't know the answer to any of those questions. To my knowledge no tests of such situations have been scientifically conducted. My own feeling is—and again it is not a black and white situation of clear alternatives—that (a) it is most advantageous to evaluate the thing *on its own terms*—in this case that includes among other items the knowledge of Strauss's intentions of writing an orchestral Rondo in his own then developing musical language, based in some both general and specific ways on selected episodes from the Till Eulenspiegel legend; that (b) a listener deprived of such prior knowledge about the work might still enjoy it immensely on its own auditory terms, but would never in a million years guess that it had anything to do with Till Eulenspiegel; and that (c) a person having this prior information, including the title itself, is forever deprived of listening to the work on purely auditory, acoustical terms. Since one cannot be conscious of "nothing," the persons in (b) and (c) are equivalent, although in different and somewhat opposite situations. Their judgments are directed not to what they don't know but what they *do* know, right or wrong, complete or incomplete, relevant or irrelevant—and are determined perforce by those variables. If they choose to exercise all their options of judgment, there then is a precise relationship between what they know and don't know, what they remember, what they intuit out of their cumulative experiences or do not intuit.

Staying with *Till Eulenspiegel* for a moment, but looking at evaluation not from the point of view of hearing it performed but of performing it, I submit that a performance which is not aware of Strauss's intended scenario and basic story line is to that extent a deficient performance, no matter how technically perfect it may be in the realization. Why? Because it does not take into account in its realization an essential part of the intelligence and the emotional impulse that created the work in the first instance.

Now if the *evaluator* does not know how Strauss's tone poem is based on selected aspects of the Eulenspiegel legend, then he, too, cannot make a proper qualitative evaluation of such a performance. The trouble is, of course, a person will not suspend judgment because he does not know something. For in the very condition of not knowing it, he cannot "bracket out," as you phenomenologists would say, that not-knowing. Similarly, if he knows that he doesn't know something, he is already well on the way to knowing it.

The hard realities here are that—despite Strauss's foolish claims to the contrary—an F sharp or a chord or a chord progression or a melody, or whatever musical element, cannot in and of itself describe anything—without the listener being told *beforehand* that it is intended to describe that "anything." Therein, of course, lies the extraordinary power of music: that by being able to say nothing precisely and absolutely, and not the same thing to different people, *it can say everything*. It is our purest and most abstract language. It is unencumbered by specific, unequivocally identifiable associations. It can at best express general states of mind, moods, very broadly generalized feeling. An F sharp is an F sharp is an F sharp, as Gertrude Stein might have put it. It cannot describe a fork or a bed or a person. And that means, in turn, that a given piece can produce a thousand differing reactions and elicit a thousand imagined associations amongst a thousand people. And they may all be valid. On the other hand, they in all likelihood could not be elevated to the exalted level of "qualitative evaluation."

We musicians—artists in general—are reluctant to use such terms as "empirical evidence." Our art is at its best too elusive for that; it is capturable by hard evidence only in its lesser and less interesting aspects. Nevertheless, just as the moth is eternally attracted to the flame and salmon eternally swarm upstream, we—musicians or teachers—must forever strive to capture those tantalizingly elusive and unquantifiable measurements against which we may then conduct appropriate evaluations. It is a never-ending struggle in which, the more we know, the more we recognize how little we know.

Let me say a few words here about comparative judgments as opposed to purely aesthetic judgments. I see them as both useful but with different although sometimes overlapping purposes and values. While I believe in the ideal of aesthetic judgments based on the thing itself—and such judgments must always be my primary and final goal—I see no reason not to use comparative judgments to arrive at those final aesthetic ones. Comparative judgments are a handy device for keeping things in perspective. Indeed, since our judgments must perforce be made on the basis of an accumulation of knowledge, viewed both historically and experientially, comparative judgments must be used as way stations in the process of arriving at evaluative aesthetic judgments. Everything is indeed what it is and not another thing. But it is precisely those other things that allow us to see the uniqueness of that one thing.

I have spoken very little about such elements in education as a curriculum. It is really too complex a subject to be treated within the scope and time constraints of this lecture. Suffice it to say that we know much more than we allow our four-year impacted curriculum to produce, and our problems are primarily political and economic. As a result, our entire field—music—is still considered somewhat frivolous and peripheral and therefore not yet a proper subject for serious research, study, and evaluation. This symposium is in its small limited way highly exceptional, perhaps unique.

So our discussion in a sense turns on itself. Like a dog biting his tail, we always come full circle—in a complex, no longer very clear pattern of cause and effect—to the same point in the circular discourse. Good strong education is the one possible answer, but the "education" to which we might turn is at the moment itself so flawed that it cannot do the job which it should be doing.

In this connection I cannot resist a critical blast at the mania for *quantification* in our modern society. We trust nothing but numbers any more. In education we do not trust the content of a thing, the substance. We quantify it, numericalize it in some primitive way—not much higher than the Nielsen ratings in television in intent or content. And all that is particularly lamentable in the arts, for they are the least quantifiable of all of humans' endeavors and strivings.

I will close with a somewhat heretical statement, perhaps even an unpopular and dangerous one: I'll accept a flawed judgment over none at all; but I prefer an individual judgment to that of a committee, with its inherent tendency to flatten decisions out to a common denominator. "Democracy" in music doesn't work. The concept of majority rule is basically anti-creative, by definition anti-individualistic. Bach, Beethoven, Brahms, or Stravinsky did not create by common consent or committee vote. Neither did Caruso or Casals or Furtwängler or Mitropoulos or Michelangelo. In the absence of absolutes and in the knowledge that unequivocal, perfected standards of decision and evaluation cannot be achieved, I would rather take my chances with some form of benign dictatorship. You can always argue against it, oppose it, and try to dislodge it. But at least it is something to depose. And it is often enlightened. The present condition, in which a great deal of mediocre evaluation bolsters and propagates mediocre artistry and standards of training, is more insidious and dangerous. In effect it virtually precludes *true* high-level qualitative evaluation.

32

The State of Our Art

A searching, outspoken analysis of the plight of good musics (of all kinds) and the fine arts in general, threatened by an environment becoming increasingly commercialized and synthetic, strangled by the appeal of mass consumption. The two-part article originally appeared in Keynote Magazine, *published by radio station WNCN in New York, in 1982, under the editorship of Sedgwick Clark.*

I. A Stranglehold on the Arts

We who labor in various artistic vineyards, tend to ascribe most of the ills and troubles we confront to economics and financial causes. We seem to be addicted to the notion that if we only had more money, more financial support for our activities—whether in music or in dance, in the visual arts, in the theater, poetry, literature, cinema, whatever—most of the obstacles and problems we face as artists would simply vanish. In our belief that better economic support will cure most of our ailments, we invent all kinds of artificial constructs—arts endowments, foundations, specialized societies and organizations, institutes, private and business "support-the-arts" groups—whose primary and often only purpose is to financially shore up the activities of a favorite artistic cause: contemporary music, chamber music, modern dance, theater, and independent (i.e., "non-commercial") cinema, etc.

Quite apart from the fact that the money-will-solve-all-the-problems theory isn't really tenable—for example, the greatest flourishing of the arts in the United States occurred during the Great Depression—depicting our problems as primarily financial is in itself a major part of the problem, since it is not even remotely the *cause,* but rather an *effect* somewhat further down the line. The real root cause of all that we as artists and arts consumers (audiences) bemoan as frustrating and inadequate in our culture is the almost total defaulting of our educational environment and the resultant stranglehold in which the commercial interests have been able to hold the fine arts for decades.

Before I pursue this theme further, a few definitions are in order, lest additional misunderstandings arise. Point no. 1: Limiting my sights now to the field of music (for you are readers of a journal devoted to music), I want to make it clear that when I use the term "fine music," I mean *all* creative quality musics, not just "classical," "symphonic," "serious," and "opera," but quality musics in all categories and all eras. This would include, for example, creative jazz, whether modern or traditional, great folk/vernacular/ethnic musics such as (in our own country) bluegrass, cajun, country fiddle, Pueblo Indian or (from other parts of the world) African or Indian musical traditions, Japanese Gagaku, Inca musics from South America, Uzbek folk music, the remarkable choral traditions of Georgia (in the Caucasus), plus a thousand and one other forms of quality music that populate this globe of ours, *as well as* various classical musics, including Renaissance, baroque, medieval, and contemporary. By the same token I *do* wish to separate those musics from the commercial musics that engulf and pollute our cultural environment: pop, rock, top-10, top-40, Nashville—whatever their various current names and marketing labels may be.

Point no. 2: I also want to clarify that my argument here is not the old and much bandied about one of "elitism versus populism." That argument is—at least as it is usually exploited—specious and based on totally false assumptions. For it is usually used to pit the "disadvantaged" against the "advantaged," and attempts to make people who cherish quality music look like autocrats, snobs, and eggheads, insinuating that there is something anti-democratic and un-American about considering Duke Ellington superior to the Plasmatics or Mozart greater than John Lennon. Indeed the elitism-versus-populism argument is completely mythical, a polemical sleight-of-words invented by clever ignoramuses to becloud the real issue: namely, that quality, creativity, and high craftsmanship can and do exist in all forms and categories of music. So can and do their opposites: mediocrity and absence of quality. *No* form of musical expression is either inherently blessed with quality or intrinsically devoid of it.

The anti-elitists obscure the real issues further by invoking audience statistics and sales figures, as if mass consumption and mass appeal automatically equate with quality. Not content with the fact that certain commercial/popular musics already attract tens of millions of buyers (consumers), these apostles of populism feel put upon if some of us occasionally dare to suggest that most of this music is trivial *as music* and not of lasting relevance. I personally would prefer to apply the terms "populism" and "popular" to all kinds of folk/ethnic/vernacular musics, but I can't do that anymore because "popular" and "pop" have come to mean that which makes money—gobs of it. Incidentally, for some people Beethoven is "popular," but that is certainly not what network television and AM radio understand by "popular." The same can be said, alas, of such great indigenous American musics as jazz and bluegrass.

What is most insidious about this argument is its patronizing insinuation that the poor and socially disadvantaged represent a "populist" culture which the presumably wealthier "elitists" look down upon. While it is true that there are cultural snobs who regard any but the classiest art forms as beneath their dignity, *they* are not the problem. Nor is it correct to assume that economically disadvantaged minorities are culturally disadvantaged. Quite the contrary. Moreover, populists constantly confuse commerical musics, which may be quite inane, with vernacular musics that may represent ethnic or folk traditions of great beauty and artistic integrity. We live in a time and a country of cultural pluralism. There isn't only *one* good music; there are many. One may like only one, but there *are* many. And all of these musics, which coexist (and even cohabit on occasion), are capable of high-quality inspired art—*and* low-quality inferior art.

If we understand these just-stated ground rules of definition we can return to the main point: namely—and now I'd like to put it more positively—that almost all of our problems could be solved by a higher level of education in matters cultural (musical, in our case). The truth is that most average Americans haven't the remotest chance of encountering quality music, quality art in their lives. That fact results in turn in widespread cultural illiteracy and, in turn again, to a serious lack of grass-roots support for quality arts institutions and activities of all stripes and categories.

Many readers of this magazine will undoubtedly regard such statements as outlandish and heretical. You, after all, appreciate good music, otherwise you wouldn't even be reading *Keynote;* maybe you even appreciate many *different* kinds of good music. You are apt to be quite complacent in your feelings that you represent a solid constituency for "good music," that you have friends who feel likewise, who frequent the opera and the Metropolitan Museum, French films at the Paris Cinema; you watch public television, and even enjoy an occasional evening out "slumming" to the music of Stan Getz or Marian McPartland or Keith Jarrett.

What you are not apt to realize is that you represent a *very* tiny minority in our society—somewhere around 3 to 5 percent. The other 95 to 97 percent have never heard of Beethoven or Verdi or Charlie Parker or Duke Ellington or Elliott Carter, do not know such things exist, and couldn't care less. The real trouble with our situation is that these 95 percent are totally unreachable by us. They live in another world, totally unrelated to anything I am talking about in this article. Even this article is directed at the wrong audience. *You* don't need this lesson, except perhaps to shake you out of your complacency. It is once again a case of the preacher preaching to the converted. I cannot reach those who could and would make the difference in our cultural ambience. But I don't exist for them, and I have no means, mechanisms, or media by which to reach them. Even if I could get my message onto Public Television, *they* wouldn't hear it, because they're watching sitcom shows, Barbara Mandrell, and *Solid Gold*.

Nevertheless it is good for us, the already converted, to understand how serious is the plight of good musics (I insist on the plural). For if we can correctly diagnose the illness, we may eventually help to find the cure.

In a culturally pluralistic society such as ours, a colorful diversity of musical tastes is quite natural. Some people love symphonic music but hate jazz; others love opera but are bored by chamber music; others live entirely on a diet of rock and roll, AM radio, and top-10 and wouldn't be caught dead in an opera house; still others respond only to dixieland jazz (the cruder the better) and find all other music boring and irrelevant; and so on in endless variations.

Now in a culturally pluralistic society, such a diverse array of likes and dislikes would not in itself be bad, *if* those likes and dislikes were based on a free choice and if the commercial musics didn't have such a total stranglehold on our people.

The sad truth is that the majority of Americans never has a chance to make a choice, to make a contact with the myriad variety that comprises the full available musical spectrum. The commercial-music establishment makes sure of that! With network television as its henchman and primary distributor, the commercial establishment dominates and determines the musical tastes of the vast majority of Americans.

And it does this extremely well, because it has learned its lessons well from American business and industry. It creates and determines for the American musical consumer how much and what will be sold.

If you doubt my words, or think I am exaggerating, ponder this question for a moment. Why is it that 215 million Americans' musical diet consists of rock and pop, while only 10 million are able to enjoy other kinds of music as well? Is it because rock and pop are better musics than Beethoven or Ellington and those 215 million are driven to top-40 music by its superior quality? The answer is obvious. The answer has nothing to do with quality. In our consumerist society, what sells is deemed good—and that to me is the ultimate cultural debasement.

And the answer tells us something else terribly important: to what total extent our educational environment has failed us. I said educational *environment* on purpose, for I don't mean only education *per se,* and I don't mean music education as such, but the whole educational fabric of our society, formal educational and informal, as it permeates our societal environment. Right now, not one of the three places where "educating" takes place has any but the most minimal and accidental relationship to quality musics of all kinds (including, by the way, *good* rock and roll). Those three places are: (1) the school (I'm targeting here especially the primary and secondary public schools, a virtual musical wasteland); (2) the home, which is also the product of the same abysmal non-education; and (3) the general leisure environment, whose primary, ubiquitous educating force is, of course, network television and its satellite, AM-commercial radio.

We see here the devastating influence of our modern, technologically "sophisticated" world, the result of the good, affluent life that comes with electricity, radio and television, amplified guitars, canned music, robots, computers, video games, fast food, leisure time, TV preachers and TV morality. Thus we inherit "music of the electronic age" and "platinum" records, music as a billion-dollar industry, music as mass entertainment, music as promotion and hype—in short, a culture where, whether in music or theater or literature or whatever, selling easily saleable, pre-packaged, "products" is all that counts and thus determines what in fact is available.

In the meantime the Office of Education spends less than one-half percent of its budget on the arts, of which music is but one. So divide that less-than-one-half percent by 8 or 12, depending on how many art forms you want to include, and you get an idea of how infinitesimal is the presence of quality music in the public primary and secondary education sector.

We have here an essentially victimized American population whose freedom of choice in matters musical is virtually denied them by, on the one hand, the failure of education and, on the other, the omnipresence of the commercial/popular-music establishment, voracious in its appetite and greed. Not that the classical-music establishment is much help either. With its operatic and conductorial jet-age superstars constantly reinforcing the impression that "classical" music began with Mozart and ended with the *Firebird* and Mahler's Fifth, it segregates itself from even smaller musical minorities who know that there *is* music before Mozart and after 1910. These minorities know that the thousands of ethnic and folk musics that populate this globe—in Asia, in Africa, in the South Seas, in the other Americas, on our own continent, some of them four to five thousand years old—are *not* primitive (as they are so often deemed by our arrogant, snobbish, establishment taste-makers) but beautiful, strong, meaningful, profound musics that collectively form our global heritage, our total musical environment.

I am not a musical dictator, I do not wish to tell people what they should like or not like. I only wish for a condition where *everyone* can make his or her choice, and not out of ignorance but of knowledge. That is the true meaning of democracy and how it best works: decisions made from wisdom, not ignorance or apathy. And that is how music relates to democracy in a free society.

Unfortunately we have little patience with our ethnic minorities and their musics; at best we ignore them to death, or, worse yet, we integrate them and, so to speak, "buy them up," until they are a part of us and our entrenched establishments.

This point is most relevant for minorities—be they black, hispanic, oriental, Indian, Greek, Armenian, Scandinavian, or what have you—as they are anything but culturally disadvantaged. They have their own fine musical traditions and cultures, many of them venerable in years and magnificent in quality. They are kept, however—in a prejudiced society—mostly invisible and inaudible.

We don't notice these musics much—except when and until they somehow become commercially attractive. Then we appropriate them, we commercialize them, we label and package them, and then we sell 'em. By God, *do we sell them!*

The vast majority of white Americans does not know, for example, that the popular musics of the United States, both past and present—and note well that today the popular music of the U.S. is also the *world's* popular music—for 150 years have all been of black or Afro-American origin. First it was minstrel music in the mid-19th century; then, in the 1890's to about the time of World War I, ragtime (a black music); then jazz, which in its white, commercialized garb became Swing; and most recently rock music, which is nothing but the black people's Rhythm and Blues, taken over and marketed to a mass public. Was Elvis Presley, who plagiarized Chuck Berry, Fats Domino, and untold black Blues singers, really a better musician than Chuck Berry is—still today? Presley got the royalty checks; Berry didn't.

No, I'm afraid we don't treat our minority cultures very kindly, even when we recognize them.

The commercial, made-for-profit musics have, of course, moved into the vacuum left by this massive failure of the whole educational environment—moved in big, and with each month, each year become more firmly entrenched in our culture and in our collective ears.

In the meantime we have developed, despite that, a magnificent national level of training in music, especially in the realm of higher education—conservatories and university music schools. But, having thus trained hordes of superior musicians, we have neglected almost entirely the problem of building a commensurate audience for all those fine musicians to play to. The valiant efforts of Young Audiences, of the Artists in the Schools, of the Endowment and a number of similar ventures, have been woefully inadequate, simply because they are hopelessly outnumbered and up against a much better organized and powerful enemy: the commercial music interests, who *do* have the nation's ear, who have a stranglehold on us.

I realize that in a period of Reaganomics and new Federalism, the arts are in an even more precarious situation, and therefore many may regard this as the worst time to rectify the situation—although the example I cited of the arts in the 1930s Depression ought to give us encouragement and hope. But, in truth, there is never a right or easy time for such things. There are always enough pressures on us and enough ostensible reasons for doing nothing. And the problem of treating the fine arts as a "fringe, peripheral hobby of a few elitists" has festered in our society far too long. The cycle of non-education and non-contact with the arts continues generation after generation. We must break into the cycle—somehow, sometime soon.

How that may be accomplished—and of course there are no easy solutions—will be suggested in the sequel to this article. Tune in next month!

II. Loosening the Stranglehold on the Arts

The utter failure of our educational system in matters cultural is one problem; the other is the almost total stranglehold "commercial" music has on our musical environment. It is difficult to escape this ubiquitous made-solely-for-profit music, aimed with such great skill and accuracy at the lowest cultural common denominator to be found in our society. It is a musical mass product which has found it easy to move into the vacuum left by the cultural illiteracy spawned by our primary and secondary schools and the commercial media. The problems we have thus inherited are of awesome proportions, with no easy answers.

But surely one answer must lie in the mobilization of a counterforce—or various counterforces—to right the serious imbalances and inequities in our musical culture. Before one can cure an ill one must be able to diagnose it, and, as I suggested in last month's *Keynote*, there are three major arenas in which this particular educational cure must be sought: (a) the schools (formal education), (b) the home, and (c) television and radio (the latter two *informal* education).

The area of formal education most in need of attention is public education, specifically in the primary and secondary levels. Higher education in colleges, universities, and music conservatories provides a broad range of excellent opportunities to gain a degree of musical literacy. Various kinds of survey courses, electives, and, of course, specialized training in fine arts departments, music schools, and liberal arts colleges offer a wide enough sampling of the world of arts to satisfy anyone's curiosity or interest in matters cultural. The problem is, however, that such curiosity and interest are totally lacking in a vast majority of higher education-bound Americans. The fact is that if fine music (or fine arts of any kind) has not become a part of a person's life by the mid-teen years, it will only in rare cases be there in adult years.

Innumerable studies have confirmed again and again that the component structure of the human being—physical, intellectual, emotional, psychological—is in essence fully formed by the early teen years. Anything not present in that structure by then is not likely to be there later. That basic structure may be refined, polished, and it may mature or broaden peripherally; but it will in all likelihood not change or expand dramatically. What that means is that if, for example, fine music—of whatever kind—has not appeared in a person's life by the age of 12 or so, it will probably not be an integral part of that person's later life. The contacts and values implanted in the early years, from the cradle to the young teen years, are the only ones that really stick. And that includes one of the most essential qualities we humans possess, one which uniquely distinguishes us from the rest of the animal kingdom: intellectual curiosity.

The sad reality is that music and the other arts have been appallingly short-changed by public education and that as a result the vast majority of Americans are the victims of a defaulted education, at least in matters cultural/artistic. It is hardly a secret that American public education is at an all-time nadir. Every day an article or survey appears somewhere deploring the failure of primary and secondary education, even in the so-called basics. But that failure is as nothing compared with its failure in music and the other arts. In some fortunate places—not necessarily big-city cultural centers—music programs are excellent, well ensconced and even impervious to cutbacks in less favorable economic times. But these are exceptions. In most school systems music is either nonexistent or treated as a "social activity" (the football and marching band might serve as an example) supported at best by a "lip service" attitude or a somewhat embarrassed nod toward music as something to be merely tolerated. Many school boards, teachers, administrators, and parents—themselves the victims of the defaulting educational process I have described—not surprisingly care little about the arts in their school(s), because they are not an integral part of their lives either.

In the meantime the stimulation and leadership that might be expected to come from the Office of Education, which funds public education in the United States through vast multibillion-dollar amounts appropriated annually by the Congress, is totally lacking. Indeed the amount of its budget allocated to arts programs in the public schools, noted last month, is so shocking that it bears reiteration here: less than one-half percent. That means that music, one of seven or eight arts disciplines, receives something like one-sixteenth of a percent—a national disgrace.

Anything to do with public education is in our day and age instantly economic and political, whether at the federal, state, or local level. Therefore changes and corrections of attitude have to be fought up and down the entire political/bureaucratic ladder. Questions relating to the arts and culture become *political* issues, because they are *financial* issues. And since only a tiny minority of students and parents and school officials are even dimly aware of the existence of fine musics, it is obvious that the struggle for any significant improvement in education's perception of the arts will be a difficult one.

Nonetheless, the answer to the problem is quite clear—and it is the only one in a democratic society: The arts community (in our field the music community) must mobilize itself into a persuasive force, a lobby if you will, which will have enough power and clout to force a change. It will have to become a political (and even economic) force, a tactic innumerable other sectors of American society, be they industries or religious groups, have long ago understood. We musicians—we artists—have stood too long, hat in hand, on the sidelines, hoping for a few crumbs of respectability and financial viability. By and large we are still perceived as freaks and aberrations in our society by a vast majority of our populace, at best bemusedly tolerated, when in fact we may be one of its best and healthiest attributes—and one of the most respected *everywhere else* in the world.

The point is—stated in all of its brutal starkness—that if we in the arts community do not organize to give culture a better place in our society, then who on earth will? Those 95 percent of Americans who don't even know we exist? Obviously not. And I must say, if we in the arts do not mobilize ourselves as more persuasive spokesmen for American culture, in all of its broad melting pot manifestations, then we surely deserve what we get.

Make no mistake about it: The public school is one of the first and priority arenas where this battle needs to be fought. For the children of today will be the adults of tomorrow, who in turn will perpetuate cultural illiteracy, in turn affecting *their* children, who then continue the process passing on cultural illiteracy to *their* children, and so on in a nauseatingly repetitive cycle. We must intersect that process. We must become vocal, articulate, active on behalf of our culture, our arts, our music; and we must not be intimidated by the illiterate majority, by bureaucratic immobility, by political entrenchments, by the pessimists in our own midst who are likely to tell us that change is impossible, is hopeless.

In the second arena, the home, the problems are similar in nature, though not at all political or even particularly economic. The issue here is essentially the same, namely that the home, i.e., the parents, must help to provide their children with a contact with the arts, or with fine music of many kinds, all through their childhood. This can be done in a great variety of ways, for example by taking children at appropriate ages to concerts, theater, fine cinema (not *just* the latest highly touted box-office thrillers), museums, photography exhibits, jazz performances (no, they are not always in dingy nightclubs), ballet, and so on. But even easier, and less expensive, is creating an artistic environment in the home upon which the child will almost inevitably feed—a home in which there are fine books, important literature (not just the latest paperback best seller and *People* magazine), a home in which good music (of all kinds) can be heard with some regularity on records and on radio (such as WNCN and National Public Radio stations), in which good drama or educational programs on the sciences and the arts are viewed with some regularity (as on Public Television or certain cable television channels).

Parents nowadays, even when they themselves appreciate fine music, are often cowed by the resistance their children show toward quality music—a resistance, of course, nurtured on the one hand by peer pressure in the schools, where children who speak correct English, love some art form, and do not spend whole afternoons playing Pac-Man are considered sissies or freaks; and nurtured on the other hand by network television and commercial radio, where murder and violence, imbecilic game shows, and top-ten musical inanities reign supreme. What many parents do not seem to realize is that it takes very little contact with fine music and other artistic expressions in order for them to take a hold in a child's consciousness. Parents forget that they have a dozen years—but especially the first seven, the most crucial in a human being's development—during which they can provide, unobtrusively and unforced, contacts with the arts. Twelve years is a long time, and it may need only one sowing for a seed to take hold and to grow into an abiding consciousness and further curiosity. For the mind and its ways are mysterious indeed. We know relatively little as to how, precisely, an imprint is made on it. But we do know that some contact—even if unnoticed and unappreciated *at the time*—may develop into a deep interest sometimes later, even years later.

Too many parents are intimidated by the omnipresence of rock and pop music in our environment, and mistakenly assume that their teenager's addiction to them is a natural preference and a matter of choice. It is not; it is rather that these musics offer no problems and challenges to the listener, and are in addition ubiquitous and unavoidable. A teenager looking for Mozart or Duke Ellington on television or radio (or in other phases of his "environment") will have to search long and hard. But parents should not be discouraged from continually trying to rebalance their children's musical diet.

For all of this is not to say that we must abolish commercial music, but rather that we must abolish its monopolistic stranglehold. There have always been and there will always be musics which are vastly more popular than others. That's as it should be, for these popular musics serve a critically important function in society, primarily as an entertainment outlet for the general public, the "common man." It is not a question of depriving him of his entertainment—indeed he would find his entertainment alternatives considerably *broadened*—but it *is* a question of giving other fine musics a fighting chance and of offering everybody, the common as well as the "uncommon" man, a choice of musical fare.

The parental/home influence cannot be overestimated. Several recent studies have confirmed that the specific environmental conditions found in the home are crucial influences in the development of young talent. Essential features of these conditions include a surrounding in which an art—music for example—is considered a natural and integral part of life, in which a healthy work ethic is fostered, and in which creative or intellectual curiosity are stimulated. These qualities need not be forced upon the children; they need only to be present on a sustained basis in the *atmosphere* of that home, permeating its environment.

The much bigger problem, of course, is that there are *millions* of American parents and homes which have no contact with fine music themselves. These homes are cultural wastelands, and cannot offer any effective counterbalance to the prevalence of *schlock* music and glitzy TV. Worse, there is perhaps no way to reach these homes. One may simply have to write them off as lost to good music, and hope that breakthroughs elsewhere can save at least their offspring.

The last arena which impacts dramatically on cultural literacy is, of course, radio and television—the prime educator for some decades now of tens of millions of Americans. The facts here are quite bluntly that fine music, fine literature and drama, fine visual art, fine *anything* simply do not occur on commercial (network) television and radio. They are instead relegated to the "cultural ghetto" of PBS, NPR, and such isolated commercial stations as WNCN, where these matters can then be roundly ignored by all but a small minority of our populace. Between the soap operas during the day, murder and mayhem in prime evening time, and the talk shows—which only very rarely and then apologetically offer anything remotely related to quality art—television is a cultural desert, a strident, corrupted market place where selling a product (noncultural, of course) is all that matters.

What is so disturbing about this is that commercial television has virtually complete control over the entire terrain, and (even worse as far as I'm concerned) that so many people *in* music and the arts are quite complacent about this state of affairs. They have evidently not realized that everything they complain about—from inadequate financial support to poor working conditions and a host of other inequities—is in the last analysis all traceable to the central fact that music and the other arts are considered in our society at best mere entertainment, a social adornment, a peripheral frill, and at worst something to be ignored, ridiculed, and abhorred.

People in the arts tend to be content with things as they are as long as they themselves have some kind of job and there is no direct threat to *their* security, as long as *they* can enjoy the pleasures and benefits afforded by the arts, and as long as *they* can watch their public television and listen to their favorite symphony or jazz program. But that is extraordinarily selfish. For it fails to appreciate the dilemma in which a few hundred million other Americans find themselves—through no fault of their own—deprived of those very same pleasures and benefits. To not care about these issues, to not care whether one's fellow human beings can enjoy a Mozart or Beethoven symphony or a Charlie Parker or Thelonious Monk improvisation or some authentic bluegrass fiddling or whatever, is to not really care about the arts and their rightful place in our culture; it is to run the risk of the arts facing an even greater struggle for survival.

In this fragile balance, television—the greatest educative medium ever invented by man—could play an enormously crucial role. But it is not permitted to do so because, as far as music goes, television has been completely captured by one kind of music, leaving room for no others, be they "classical," vernacular, popular, or ethnic. It is clear to me that the statistics I gave last month—five percent of our population aware of fine arts, 95 percent unaware—will never change until commercial television and radio are pressured into allowing culture onto the public airwaves. Right now, the networks feel no obligation to present anything of quality and creativity, no obligation to inform or elevate the public. Network executives blithely invoke the specious rationale that they are only giving the public what it wants. But the public's self-perpetuating cultural illiteracy is further fostered and exploited by the networks, when even a modest amount of arts programing could have a substantial impact on the whole populace and our whole culture.

For disbelievers there exists already one striking example of how television might positively affect and support the arts. WBZ-TV in Boston, an NBC affiliate and Westinghouse station, embarked late last year on a campaign to support the arts in Massachusetts. Called "You Gotta Have Arts," it consists of an actual fund-raising campaign (with a goal of $1,000,000 in one year) and significantly stepped-up programing in and reporting on the arts, mostly in prime time. As a remarkable experiment, unprecedented in the history of commercial television, it is already proving to be amazingly effective and, if duplicated in only a dozen other metropolitan centers around the country, could have a dramatic impact on the place of the arts in our society.

None of these things will happen, none of these possibilities will be realized, unless we in the arts and the intellectual community provide some leadership in directing our society toward such goals. But first we must shake ourselves loose from our complacency and recognize that, as good as our musical culture is, our country's talent potential could make it even much better. What we need now—in the field of music, at least—is not more artists, more musicians, but *an audience* that can support our musical institutions, performers, and creative artists and, over the years, cause our musical culture to flourish and to become a part of *all* Americans' lives, not just a lucky few.

33

Form, Content, and Symbol

A talk delivered in the early 1960s in Monterey, California, as part of a series of interdisciplinary symposia on the arts. Schuller here analyzes the complex and often misunderstood relationship between form and content and, beyond that, the thorny subject of the potential meaning *of music.*

RECENTLY, IN CONNECTION WITH a Symposium held to discover what (if any) similarities or parallels existed between the various arts, it was brought home to me once again in a very direct manner that music is indeed the most "abstract," the most "absolute," the most non-representational of all the art forms. It also confirmed for me the long-standing impression that the non-musician, the non-professional music lover, and lay observer have a tendency, understandable enough, to relate their musical experiences to specific emotional events or concrete tangible physical phenomena or extra-musical associations. Each looks upon these as aids to an appreciation of music. But they are, in fact, obstacles and limitations to a full understanding of a given musical work. It seems to me, then, that my first task here is to define the essential nature of music, what music is capable of, as well as what it is *not* capable of; what its various functions are and have been throughout the history of music, and how—if at all—its external characteristics have changed in our own time.

While everyone seems to agree that music is a powerful communicative phenomenon, there is disagreement as to how and why music communicates thus, and from there—one step further—what therefore music's function is or should be. Fundamental to these differences of viewpoint has been the question: Is music basically a reflection of life (or more accurately, man's view of life)—or is it a domain unto itself? Should the composer direct his music at the feelings and emotions of man, at his soul—as the philosophers would have it—or should he preserve the purity of music by rejecting all extra-musical elements, constructing autonomous structures of "organized musical sounds"?

Since ancient times these two contrary positions have been the source of continuous philosophical arguments and discussions, without either side ever emerging victorious. Perhaps we will find that "the truth" lies somewhere between the two extremes, and that, if at certain periods in the history of music one view seemed to be winning, sooner or later a reaction set in and brought the two opposing tendencies back into line and into direct opposition.

The two approaches have been characterized variously as the Dionysian versus the Apollonian ideal, as emotionalism versus intellectualism, the "expressive" versus the "formal," as "absolute" versus "program" music, and so on.

There are people who contend that "absolute" music constitutes no more than a meaningless, irrelevant acoustical abstraction, that music can and should be a strictly functional art, as dance music, say, or commercial music; or that, at its noblest, it must express something specifically *representative,* a story or a poem, as in the tone-poem genre of the late nineteenth century. There are others, of course, who are quite certain that music can stand on its own two musical/acoustical legs and needs no literary, poetic, or programmatic buttressing—that a symphony, for example, does not need to rest upon a supporting story line or a specific mood or character in order to be understood and experienced as an artistic expression.

Apart from the late nineteenth-century tone-poem movement, music as a specifically descriptive, representative art actually had its great heyday in the eighteenth century, especially in France, where the imitative skills in music became highly developed and were actively taught as the *raison d'être* of music. The eighteenth-century French mathematician Jean d'Alembert said: "Toute musique qui ne peint rien, n'est que du bruit" ("Music which does not paint, is nothing but noise"). His compatriot and contemporary Denis Diderot believed he had never heard a good piece of music, especially an *adagio* or an *andante,* which he could not interpret by means of a story. In our century these ideas persist in the spread of so-called "music appreciation" classes, and in the theses of certain musicologists, including as important a figure as Arnold Schering, that every symphony of Haydn and Beethoven revealed some hidden program or literary idea. Another point of view was a little more tolerant in that it conceded the existence of non-programmatic music (like a baroque sonata) but only because it enabled the listener—so the reasoning went—to give full vent to his imagination and to create any number of "fantasies" and "visions" relating to or inspired by music. The underlying assumption was still that only a story or program could make this music meaningful.

Diametrically opposed to such attitudes was the viewpoint of the famous late nineteenth-century Viennese critic and intellectual Eduard Hanslick, whose fame now rests primarily on fanning the flames of the feud between Brahms and Wagner. Hanslick, one of the first propagandists for "absolute music," stated without allowing for any qualifications that "the true work of art in music has no other purpose than to be itself, and has no other content than sounds formed in time." He rejected emotional expression in music. Ironically, he succumbed to the very emotionalism he tried to exorcise from music in his own dialectics on the subject, a weakness which led him to state such extravagant theories as, for example, that expressive music was "pathological," while absolute music was "aesthetic."

A moderate and more middle-of-the-road approach was taken by Schopenhauer, who said that "music cannot express a specific sorrow or a specific state of happiness, but sorrow in general and happiness in general, i.e., these feelings *in their essence* without any specifics and without any specific motivations or mundane relatedness to life." Here Schopenhauer foresaw in music "a revelation of the innermost essence of life." Even a man like Wagner, who in most ways thought of his music as a servant of his philosophy, sometimes took a more moderate position and admitted that Mozart was a magnificent musician, even though his creations were primarily examples of "pure music"; and he spoke even of Beethoven's "primeval and absolute tonal language."

Such varied shadings of opinion regarding the existence or nonexistence of "absolute music" have their counterpart in the music itself, where we find that there are infinite gradations of "absoluteness"—a factor which has contributed to the controversy between the Apollonian and Dionysian viewpoints.

"Absolute" means absolved, closed off, independent, disengaged, and pure—in our context: the separation of music from its worldly environment, that is, non-functional music (in the sense of functional *Gebrauchsmusik* or descriptive theater music, or the like). Now, instrumental music does not really divide readily into two sharply separated zones—absolute music and program music—but rather must be partitioned into many gradations and overlapping categories, located somewhere *between* these aesthetic poles.

In this connection we can observe some interesting analogies to the other art forms. In painting, for example, there exists a series of categories and styles, running the gamut from photographic representation to complete abstraction. Likewise in music, there exists great variety of styles ranging from an almost phonographic reproduction of an event in nature, like the chirping of birds (as in Strauss, or in our day, the French composer Olivier Messiaen) to an extreme purism and asceticism (as in certain baroque instrumental fugues, or the piano music of Erik Satie). In addition, we can differentiate between *kinds* of absoluteness: from *what* is music absolved? Is it absolute and free from words, objects, or people? Also: whether a piece of music *consciously* determines to remove itself from such extra-musical associations, or whether it is already *sui generis,* intrinsically opposed to such associations? (The former, for instance, might be an opera or an oratorio attempting to be abstract, like the German composer Boris Blacher's *Abstract Opera;* while the latter could be a fugue or an instrumental sonata which *can* have extra-musical associations, but only if the composer (or the listener) wishes to attach such associations to the work in question.)

In any case, it can be seen from these examples that the alternative concepts of absolute and program music—in essence musically versus extra-musically related, autonomous versus heteronomous—are simplifications. And the history of music can show us some curious and paradoxical examples thereof. On the one hand, people have attempted in every era to discover in music a greater degree of expression and symbolism—we think of the religious and mystical symbolism of the Middle Ages and Renaissance, the pictorial associations in much late Renaissance and early Baroque music, the attempts at realism in opera, especially in the Italian "verismo" style, and so on. On the other hand, simultaneously there has always existed a strong tendency to discover the purest, absolute musical "essence" of music. Random examples of this are not only a work like Bach's *Art of the Fugue,* but the many instances of the same text being set to different musics or, vice versa, the same music exploring different texts; similarly Wagner's use of a certain *leitmotif* in totally different plot situations. Or take a modern extreme of this manifestation: certain moments in the works by Charles Ives which were never intended to be performed because they defy accurate and complete performance realization;

they are musico-structural abstractions, intended (or doomed) to exist simply as musical *ideas,* to be heard in the minds of certain receptive musicians who are capable of reading them in notation and hearing them in their inner ear. It would be difficult to get much more "absolute" than that.

So we can see this dualistic approach—the "absolute" versus the "programmatic" and "functional"—from Josquin des Près to Schoenberg and beyond. On the one hand, gradually proliferating techniques and methods permitted an increasing range of expression of human, spiritual (or even demonic) values. At the same time, another branch of music carefully used its techniques to achieve what one might call an absolute absolutism. I think this conflict and the contradiction of two opposing tendencies are a logical outgrowth of a similar dualism in life itself, found in man's everyday existence. Man is continually drawn toward viewing, examining, dissecting, investigating, and specifying life and his environment, and at the same time apparently wanting to also escape from his surrounding realities and to transcend them by whatever means are available to him: artistic, spiritual, religious, or otherwise.

Another phenomenon (and paradox) of music is the fact that a single tone or a chord or a melody literally represents nothing but itself, a specific sound, a series of acoustical vibrations, which in addition must be reproduced in a *performance* before it can have any impact or effect at all, something not true of a painting or a book or a film. On the other hand, music is a limitable, prescribable mode of expression and communication. One need think only of the many specific functions musical sounds have been made to serve since the beginnings of man. Music is used as listening and dance entertainment, as all manner of background for other art forms (theatre and film especially), even in restaurants and cafés, as well as in churches, in ceremonies and functions, even as a means of selling products in advertising, etc.

In fact, our Western "absolute" music (in the sense that Hanslick used the term) is, if we take the world's musics as a whole, a historical and sociological exception. In most cultures, music is still primarily thought of as functional or ritualistic, as a means of producing certain desired effects upon human beings. The ancient association of music and magic comes to mind here. Music in such a context could obviously not be a thing unto itself, but a means to an end. It is in this sense—music as a means of creating moods and feelings in the beholder—that music has most often been used, even in Western culture.

But, as mentioned, this is mostly an illusory conception of music. Music is capable of reflecting or expressing happiness or sorrow or similar human conditions, however, always in the *non*-specific sense. For it must be understood that the making and performing of music are artistic, stylized, humanly developed skills—quite separate from the pure existence of sounds in nature, for example. The elements of music coincide only partially with those of the real concrete world. This is why music cannot automatically and unequivocally describe reality, at least not as finitely as can painting or literature, especially in its drama or novel forms. Music cannot specifically reproduce the realities of a story, say; it cannot reproduce for us a place, a building, or a person, or a specific act or time. Even as naturalistic a piece as the storm music of Beethoven's "Pastorale" Symphony is worlds removed from the actual realities of a true storm. It is a remarkable evocation of a storm, and I suppose it is one of the rare moments in music that even an uninitiated listener might recognize immediately as a kind of storm music. But still, it is several times removed from reality, and at most recreates *some of the sounds* of a storm.

The most that music can accomplish is to abstract the qualitative essence of something in general terms, without being able to represent the specifics of such a situation. Music certainly cannot depict objects in their full, unmistakable concreteness and individuality (as Richard Strauss once in a boastful moment claimed, when he said that he could even depict a knife and fork in music).

Even when a text or story is used in music, it is the unique power of music that denaturalizes that text or that story. This capacity is exemplified clearly in the use of word or phrase repetitions in music, as in the *Da Capo* arias of Bach, say, or even in so special a factor as the relative unintelligibility of most sung texts and librettos. How often have we enjoyed vocal music despite the fact that we did not understand a word of what was sung, either because the singer was incapable of enunciating clearly or because we did not understand the language in which the music was being sung. A great Puccini operatic aria, for example, sung out of context in a concert and heard by an American listener, shows us at once the unique, overwhelming communicative power of music, but also its inability to recreate specific conditions and situations, without in either case necessarily lessening our enjoyment of the music.

The most common kinds of absolute music—or as I have tried to show, *relatively* absolute music—have been certain forms of instrumental music, such as fugues, canons, sonatas, preludes, etc. In these, the composer generally does not give more than a tempo indication. Sometimes he will add another marking, such as *mesto* (sorrowful), *con amore, con passione,* or (in German) *sehnsuchtvoll* (with longing) or *heiter* (lively), and so on. Such terms are used to help the performing artist to properly re-create the mood and character of a given piece. In fact, without such character indications most "romantic" nineteenth-century music is quite unperformable. Naturally—and interestingly—the character of a piece is present the moment the composer hears the notes in his mind and writes them down. However, due to the relative inadequacies of our musical notation, the character and mood of a piece of music often cannot be *fully* expressed by notation alone. It needs some descriptive clues, and, as already mentioned, it needs, of course, an acoustical realization.

In this connection, many musical forms arose precisely to fill the need for a particular kind of musical expression, which with further usage was in turn automatically expected to be heard in that form. I'm thinking, for instance, of the *Sarabande* in Bach's solo suites, or the *Berceuse* (a lullaby) in Chopin's music—or his *Nocturnes* for that matter. But again, I must come back to the point that the implications of these titles or descriptive adjectives remain very nonspecific. There can be all kinds of *Vivaces:* for instance, a gay *Vivace,* as in a Haydn symphony, or a demonic one, as in a Berlioz or Liszt work. A *Nocturne* can be quiet and contemplative, or it can be turbulent and restless. So that even a title like *Nocturne* can only be taken in its loosest and most general sense. To compound the problem and to underline once more the inability of music to be unequivocally specific (so that *all* people would understand exactly the *same* thing from a given piece of music), the actual performances and interpretations of the two types of *Nocturnes* just cited, might obscure the composer's original intentions. And this is not at all exceptional. In fact, the misrendering of music by performers occurs frequently enough. One performer of our hypothetical "turbulent" *Nocturne* might, for example, feel it in a more contemplative vein than the composer intended, and vice versa: the performer of the contemplative type of *Nocturne* might feel it more violently than intended. Except through the establishment of a tradition—a fairly rigid one at that—no one but the composer can say with complete assurance who most accurately represented his intentions. And since most composers performed, especially in our time, are not amongst the living, we cannot even have such reassurances as a possible recourse.

Through the centuries, not only did certain *forms* connote certain moods to composers, but even certain keys began to take on rather specific characteristics. This particular development occurred only after our diatonic scale was "tempered" in the eighteenth century. The premise of this theory was that certain keys and modes took on *a priori* meanings, which the music could not escape. D-minor, for example, was associated with moments of great portent, with very serious or grave music, or with the expression of certain daemonic elements. E and A major, on the other hand, were associated with happy, bright, spontaneous moods, with thoughts of spring, and renewal. It is still hotly debated whether such associations were *inherently* implied by the acoustical characteristics of these keys, or whether such associations developed *after* a particularly compelling example of music using a given key in a given context became well known and influential with other composers and listeners. For example, I imagine that Mozart's use of the key of D-minor had a great deal to do with associating that key with the portentous, the grand, and the daemonic, because those conditions, when they occur in his opera *Don Giovanni,* are most often set in the key of D-minor. When Beethoven set the first movement of his 9th Symphony in that same key, it must have seemed to some people that D-minor did indeed have a built-in sense of grandeur, all its own and not to be shared by another key. As I say, this point has never been satisfactorily, let alone irrefutably, demonstrated—especially since we now know that pitch levels have changed dramatically since Mozart's day. The A we tune to today was about a third lower in the seventeenth and eighteenth centuries. Obviously, such a change would have a great effect on the supposed *intrinsic* acoustical characteristics of a key like D-minor. In other words, the pitch we now hear as D-minor was actually F-minor in Mozart's day, but sounding a third lower corresponded to our D-minor. Similarly, Mozart's D-minor was actually our B-minor as he and his listeners in reality heard it.

In any case, a man named Paul Wies conducted some experiments with non-professional musicians, lay listeners, in trying to test the theory of the characteristic uniqueness of key centers. He played a Schubert *Impromptu* in G-flat a half tone higher (in G), and discovered that almost all his listeners preferred it intuitively in the original key of G-flat. The real significance of this experiment and statistic remains yet to be fully diagnosed and understood.

But if there is this kind of sensitivity to key centers in listeners, there is even greater sensitivity to tone colors (as produced by various instruments) or to tempo deviations. Experiments have shown that deviations from an originally intended tempo, for example, did not only seem less logical to uninitiated listeners, but to their minds seemed to result in virtual caricatures of the original. It is this capacity for changing the character of a theme by tempo alterations which led to the compositional devices known as "diminution" and "augmentation," devices by which a theme is doubled (or otherwise sped up in tempo) or else slowed down. This has often been used for comical effect, and is automatically understood as such by even the lay listener—another example of how a particular technical device, "absolute" in a sense, has a meaning beyond itself.

To delve even further into the complexities and subtleties of music and its potential meanings, there was a time—roughly before the nineteenth century—when composers were content to set a given piece of music or a movement in one single mood, one tempo, one character. But after that, with the gradually expanding means of expression available to the composer in the nineteenth and twentieth centuries, the particular emphasis placed upon a work—whether in the direction of absoluteness or of programmatic content—began to vary from style to style, country to country, and eventually even during the course of a single work. From a single more or less pre-set form and mode of expression in pre-nineteenth-century times, the resources of expression proliferated into a much broader range of possible techniques, styles, and conceptions.

In a contemporary work, for instance, it may well be that the modes of expression and their particular emphasis may pass successively through a whole range of musical elements—rhythm, melody, harmony, timbre, or whatever. A piece may start with a strong energetic rhythm, with no other connotations implied. Soon, however, the emphasis may shift to a more lyrical or emotionally expressive level, becoming less rhythmic and more involved with a *cantabile* (singing) melodic approach. A little later still, the emphasis might shift back to a greater degree of "abstraction," as a highly complex contrapuntal section develops. In another section of the work, the emphasis may be momentarily on tone color or timbral considerations. What I am describing here is, of course, a hypothetical piece, but such works abound in increasing numbers ever since Beethoven's "Eroica." One need think only of the latter movements of Berlioz' *Symphonie Fantastique*. Today, in most recent works, the shifts between such different planes and levels of emphasis occur in a much more fragmented, rapid, and elusive kind of continuity, whereas in Beethoven we still experience such shifts in fairly prolonged, easily discernible time spans. That is, in Beethoven shifts may come every twelve bars or sixteen bars, and in symmetrical structures, while in contemporary music they may occur in a few measures or even a single bar. It can even happen, as in the music of Charles Ives, that these various planes and levels do occur not successively, but *simultaneously* (vertically), inseparable from each other. These have meaning only when such a moment of music is heard in its inseparable totality.

But even the most abstract, the most absolute kind of music contains some degree of emotional expression. This is what the partisans of purely "absolute" music often fail to understand; they see music only as a series of events of greater or lesser intellectual complexity. They wish to see only the syntax or technique of a musical structure, forgetting entirely that a piece of music may have a "heart and soul" without being expressive in a *specific* way.

The difficult thing for many people to understand is that, while music cannot *literally* describe a person, or a house, or an event—as a sentence or a painting can—music may nevertheless be *informed* by such images, as a kind of background psychological, creative impulse. Curiously, a work may actually tell us more *about the composer* than the composer perhaps intended. Of course, *what* it tells us may be primarily subjective, in the realm of interpretation—ours. But there is no question that, as folk dances or folk music reveal much about the people who created it, so art music tells the keen observer much about the character of a given composer, and often the milieu, the environment in which he lived and worked.

In fact it is these differences in character which give music such a broad expressive and conceptual range, as well as such a limitless fascination. In a technique like "serial technique," one of the more recent methods of composition, often erroneously thought to be confining and constricting, it is not at all difficult to discern personal and stylistic characteristics, or even elements as general as national characteristics—at least in the case of the better composers.

It is entirely possible, for instance, without prior knowledge to hear such differences in the works of Boulez, Stockhausen, and Nono of the 1950s and 1960s, all using more or less the same technical points of departure. Allowing for a certain degree of oversimplification to make my point, it is possible to hear immediately Nono's italianate lyricism, which somehow breaks through the tortured rigors of his serial compositional technique; Stockhausen's less lyrical, more structure-oriented approach (without, of course, being inexpressive), and his rigorous, consequent, typically German turn of mind, making him emphasize theory almost as much as practice; and then Boulez, in whose music the typically Gallic characteristics of sensitivity to color and dynamic shadings—and to poetry—are easily discernible. These elements are not consciously striven for but are there by virtue of the fact that a creative artist, if he is genuinely and individually creative, cannot distance his music from himself. He cannot "absolutize" it, so to speak—even when the means employed may be considered by some completely abstracted and mechanized. The composer's humanness and individuality will break through in even the most abstract or absolute or technical/intellectual musical forms.

But music can tell us that much about itself and about its composers only at the price of not being able to tell us more. And if that sounds like a riddle, it is. Its twin capacities to mean all and yet nothing represent two sides of the same coin. In this respect music is like mathematics: it makes us aware of concepts and ideas which do not exist in physical concreteness, tangible reality. Similarly, at its highest level music allows us to see and experience spiritual states and conditions, which on the other hand some of the other arts find it difficult to evoke. A painter can much more easily portray a house or a face than a state of bliss. With music, the opposite is true. We must only guard against demanding from music too specific a rendition of these feelings. Indeed, to ask more of music than it can give is to limit one's own enjoyment of it.

I do not wish in any way here to propose (or defend) a kind of "elitist" snobbism when I suggest that the proper appreciation of music is limited to relatively few people. I say it, however, because it is a truism, a reality after years of empirical evidence through personal observation and study of the history of music as a sociological phenomenon. It is this condition which accounts for the fact that absolute music will never be "popular" in the mass sense of the term. Popular appeal will always be directed first and foremost toward such music as readily discloses an obvious association either with a text or a story, or—as in the case of certain popular instrumental pieces—an unabashed, immediately recognizable sentimentalism (Dvorak's *Humoresque,* his *Largo* from the "New World" Symphony, or Tchaikovsky's *Andante Cantabile*). Even relatively enlightened and sophisticated concert goers have trouble digesting a piece of absolute non-associative music—like a Bach fugue or Beethoven's *Grosse Fuge* or a Milton Babbitt composition—because they become used to equating art with concrete objects or tangible events of a specific nature.

I wish it were possible, by education and enlightened persuasion, to stamp out this rather lazy and accommodating way of listening to music. For it is also a risky and untenable assumption that such listening leads to the best understanding of music. For example, it can easily be proven that a tone poem or a work with specific literary connotations only begins to mean those specific things to a listener *if he is told these extramusical facts beforehand.* Experiments to this effect have been conducted often enough in recent decades, definitely establishing that the uninitiated listener to Strauss's *Till Eulenspiegel,* say, would never guess *just from listening* to that work, that *that* is the story Strauss had in mind. The composer Ernst Krenek conducted experiments of this nature in his classes in California, I believe, using works like Berlioz's *Symphonie Fantastique.* It should not surprise us that out of dozens of students who had never heard this work before, only an especially "psychic" one was able to guess that the work dealt with a man who was in a tortured mental state which nearly drove him to insanity—reasonably close to the "program" indicated by Berlioz himself. But this was a totally exceptional case.

I conducted a similar experiment about five years ago with a work of my own, written especially for children's concerts. At one of these, the conductor asked the children—ages seven to eight—after the performance to tell what associations they had had to the piece, what stories they imagined as they heard it. It will not come as a great surprise that each of the children questioned came up with an entirely different story, *all of which,* interestingly enough, were plausible in relation to the music, and about which all of the kids felt very intensely, almost as if they were creating something of their own.

Significantly, the only people who failed to get anything out of the piece were the children's teachers. (I should mention that this was a work written in the so-called atonal, twelve-tone vein.) The teachers immediately concluded, after a few bars, that this music could not possibly mean or say anything to them, since it was "dissonant" and therefore "ugly" and "incomprehensible." The kids on the other hand, not knowing atonality from tonality, were not inhibited in their enjoyment of or associations with the music.

A much more celebrated example of prior, prearranged associations of non-musical ideas with musical works, is the famous case of Disney's use of Stravinsky's *Rite of Spring* in his film *Fantasia.* As we all know, the *Rite of Spring* was originally conceived as a ballet dealing with primeval rituals in ancient Russia, associated with the coming of spring. Disney—aside from drastically cutting and mutilating Stravinsky's work—used it to tell an entirely different "story," having to do with the creation and evolution of the world—a sort of oversimplification of Darwinian theory—ending dramatically with the Age of the Dinosaurs and their decline.

There are those who state flatly that if that film made some people who would ordinarily never hear the *Rite of Spring,* even in a mutilated form, hear the work, it is a positive development. I'm not so sure. Nor was Stravinsky, who objected vehemently to Disney's and Stokowski's treatment. Potentially it gave a false impression of the *real* work. Indeed it probably *diverted* innocent listeners from the original. In a strict sense it is fair to say that the only people who could uncomplicatedly enjoy the *Fantasia* film version of Stravinsky's great masterpiece were those who were hearing it thus for the first time, and who knew nothing of Stravinsky's real intentions. Such people simply couldn't know any better, being unaware of the original work. What is so dangerous here is not only that *Fantasia* gave a wrong impression of Stravinsky's work but that, in its oversimplified commercialized way, it became *more* accessible and palatable than the original. The addition of a story line made it momentarily acceptable to lots of people who would in a concert performance have rejected it; and I am convinced that those people who needed a story line in all likelihood never got to the real work itself. If so, that is a pity, because *Rite of Spring* is one of the great unqualified masterpieces of twentieth-century music, and indeed of all time.

Schopenhauer spoke about this tendency and its dangers in no uncertain terms. He said "we have the idea that we can the better understand a piece of music if we let our imagination deck it out with flesh and blood realities, garnered from nature and from life. However, on the whole, rather than aiding us in our understanding or enjoyment of music, this approach adds a foreign, willful element,—a kind of *Ersatz* experience. It is therefore much better to receive music in its direct and pure essence." And by that, Schopenhauer was implying that it is better, if given a choice, to understand *less* of the music, than to dress it up with all manner of *extra-musical* assumptions and trappings.

One might ask: what about those cases where a composer was specifically stimulated by a poem, a painting, or a historical event? In such instances, we must remember, the original stimulus does not *per se* become the content matter of such a piece. Aside from the fact that all too often these supposed inspirations turn out to be merely alleged anecdotal "embellishments," supplied by overly eager biographers or musicologists, such stimuli must remain in the prehistory of the work, a remote background of it. To make the point in its full absurdity, the *Moonlight Sonata* by Beethoven does not contain moonlight, nor does it describe moonlight in an unequivocally definable way. It was a name appended to the work many years after Beethoven's death. Nor is the cat, who, leaping across the keyboard, is said to have inspired Scarlatti to compose his *Cat's Fugue,* the musical content and substance of that piece.

This brings us full face to the relationship of form and content in music. Perhaps, I should here include the term "technique." For in the great masterpieces of music—whether a superb Bach Fugue or Chorale, Handel's *Messiah,* Mozart's *Marriage of Figaro,* Beethoven's *Eroica,* Duke Ellington's "Koko" or "Blue Light," Stravinsky's *Rite of Spring,* Schoenberg's *Erwartung,* or Babbitt's *Composition for Twelve Instruments*—show us that ultimately technique and form/structure at their most sublime levels *are* equivalent to the content; they produce the content. One can also reverse that statement, with content as the subject: content = form/structure/technique. For the truth is that content and form occur, at least in the highest creative hands, simultaneously and concurrently. It is the confluence of *both* aspects in a moment of inspiration (or for that matter in a prolonged period of white heat inspiration, as in the twenty-three days it took Handel to write the *Messiah,* or the seventeen days it took Schoenberg to write *Erwartung*) that produces what we listeners and appreciators see as monuments of perfection and beauty.

It is only in the lesser creative efforts that the discrepancies between form/technique and content show. There the edges of either terrain are frayed; the lines of demarcation show where technique begins—and ends—and content no longer precisely matches form.

Ultimately, there is no significant breach between a perfection of technique producing a concurrent perfection of form (and expression) in Beethoven's *Eroica* or Babbitt's *Composition for Twelve Instruments*. For, what we call inspiration—the moment of inspiration, if you will—is precisely the coming together, in some mysterious alchemical way, of technique and content, of mind and soul, of intellect and emotion, of thought and instinct. At such high levels of creativity, the result the composer creates is in a sense beyond his simple control; at that level a higher form of inspiration takes over. That is the mystery of creation, a process and phenomenon we do not yet fully understand or can adequately analyze. And perhaps that is for the better. For it is sufficient, for the moment, that we human beings are at least capable of such creativity.

I am not at heart a polemicist or a dialectician. In fact, I have great distrust of either skill. I have no theories to advance or advocate here, no innovations to propagate or defend. But, as a composer and a performing musician, I am greatly concerned with the most authentic re-creation of musical works. As such I wish to merely state, as exactly and objectively as possible, what the nature of music in its infinite variety is, so that we can appreciate it better, perform it better, and that in the end, stripped of all extramusical ballast, it may take on a greater meaning for us.

Thus we have seen that, paradoxically, the more specific in meaning we make our demands for music, the less it can mean to us. Conversely, a less specific meaning can in the end reward us with a much deeper general meaning and understanding and experience.

To summarize, I see form, content, symbol, and meaning not as separate and opposing elements, but as inseparable aspects of a total symbiosis. A failure to understand this must inevitably lead to a misunderstanding of the great works of musical art of the past, and most certainly will prevent an appreciation of the music of our own time, with its new and myriad formal, expressive complexities and subtleties.

In the end music can only speak for itself. It will not allow itself to be converted, subverted, diverted into other forms and substitute expressions. It is in its pure, absolute essence the most powerful—and mysterious—communicative medium man has invented.

34

Form and Aesthetics in Twentieth-Century Music

An address delivered by Schuller in October 1980 during the sixteenth annual Nobel Conference held at Gustavus Adolphus College in St. Peter, Minnesota. (The conference is the only formal Nobel program in the world outside of Sweden and Norway.) Other speakers included William Nunn Lipscomb, Jr., winner of the Nobel Prize in Chemistry in 1976, and Chen Ning Yang, awarded the Nobel Prize in Physics in 1957. The conference's theme was "The Aesthetic Dimension of Science."

THE HONOR I HAVE* just received makes me even more apprehensive than I would normally be in such august company as I find myself on this occasion; and I must confess that it fills me with some awe and considerable trepidation to stand here in the midst of a community of scientists, scheduled to speak in a conference dedicated to the theme of "The Aesthetic Dimension of Science." This is not merely because we non-scientists tend to hold the scientific fraternity in such exalted veneration, not merely because scientists enjoy such an enviable status in our society unattainable by artists, but rather more that it is a deeply engrained belief, right or wrong, that the work and product of scientists flow from the domain of logic and thus constitute an "exact science," while the endeavors and creations of artists—musicians for example—comprise an inexact science, if a science at all. In a world in which the fruits of science and technology are considered more useful—because more readily measurable than those of the artist—we musicians find ourselves on a somewhat unequal footing.

* The Gustavus Adolphus College Fine Arts Award for his distinguished contribution to the arts.

To be sure, the prevailing wisdom that scientific inquiry and discovery are inherently more capable of objectivity, while artistic creation is essentially a subjective, unquantifiable process, may upon closer inspection turn out to be myth and illusion. But there is little in our educative and societal environment that counters or disputes this myth. And so we artists are perceived as operating in a world which in no way relates to that of science and technology.

I think the planners of this conference think otherwise, and indeed hope to discover in our discussions during these days commonalities, analogies, parallels between and among these fields that need to be better articulated and understood. Feeling much the same way, I find that my qualms are mixed with a deep sense of privilege at being here, with the hope that these discussions can help to illuminate some of the mysteries that not only surround these respective fields, but usually keep them segregated and isolated from each other.

Of course, millions of words have been written and spoken over the centuries, even millennia, about the relationship of art and science. Let me just enumerate a few examples. We all know, I'm sure, about the various theories which equate, for instance, music and mathematics via the common bridge of acoustics; or the long-held belief that music is but a branch of mathematics; or for that matter the consistent pairing of the terms "science" and "art" from Greek antiquity on to the present day; or the ancient theories, still widely held, that music is but an audible exteriorization of the vibrations of life forces themselves—"vibrations of the cosmos" and "the music of the spheres" are the catch phrases of two of these notions; or again, the countless examples in the music literature of various mathematical or numerological approaches to the art of musical composition, from Johann Sebastian Bach to Alban Berg, John Cage, and Milton Babbitt. Or the various concepts of constructivism in music, especially that of symmetrical constructions observable in the works of a number of early twentieth-century composers like Schoenberg, Scriabin, Bartök, and Charles Ives; and in our own time the application, for example, of stochastic principles and those of information theory in, let us say, the works of composers like Yannis Xenakis, who is also a sometime mathematician and architect. Or the now-prevalent reality of computer-generated music where, obviously, twentieth-century scientific and technological advances play a central role in the very creation and reproduction of music. The list could go on with other lesser and greater instances of close ties between music and science.

Often enough such ideas were regarded, at least by their authors if not by the rest of the artistic community, as major breakthroughs, as innovations that purported to be panaceas which would solve in some objectifiable way problems which had previously defied resolution. One thinks of the highly touted Schillinger system of composition and compositional analysis some decades ago; of the theories of the technologically oriented futurist movement in Italy and Switzerland around the time of World War I; or, I suppose, even the concept of Schoenberg's "method of composing with twelve tones which are only related to each other"—insofar as this concept was understood (but I think more often misunderstood) as a *system,* a mathematical system at that.

No, there hasn't been exactly a shortage of attempts to systematize music, to formularize it, to quantify it, to absolutize it, to objectify and rationalize it, almost always out of an urge to emulate science, to make out of the "science of music," as so many music theoreticians of the Renaissance loved to call it, a more exact and measurable science. And yet in nearly four thousand years of contemplation of the phenomenon we call music, from Ptolemy and Plato to Leonard Meyer and Schenker to this very day, no universally accepted or even universally understood definition of music, of its essence, of its aesthetics has been realized.

So if we speak of form and aesthetics in music and especially in twentieth-century music, as the announced title of my address suggests that I do, we really stand before a formidable task where even (as is relatively easier in the sciences) a list, an accumulation of historical data and a list of achievements—not just claimed achievements, but real achieved achievements such as scientists can point to—eludes the musician and other artists.

But that's not altogether a bad thing, because I think therein lies the extraordinary power and beauty of music. That precisely because it is as an art essentially non-utilitarian, at least in our Western civilization's concept of music and art, and because it can say nothing specific or incontrovertible, it therefore can say and be everything. Precisely because the purpose and essence of music defy unequivocal, scientifically demonstrable defining, music speaks to us in ways that are at once profoundly moving and deeply personal as well as infinitely variable and diverse.

So what we are left with—and have always been left with—are viewpoints, theories, opinions, intellectual or aesthetic positions, which are determined in turn by a multitude of factors, conditioned in lesser or greater degrees and in a staggering variety of ways by background, education, fashions, individual capacities, be they physical—in music, for example, the aural or auditory capacities—or be they intellectual. Indeed, for all we now know our viewpoints are conditioned—possibly—by genetic factors and, finally, by aesthetic and human values not directly related to music, extant in the society at large. And mixed into all of this—as if it weren't already complicated enough—is something we in music dare to call "progress." For we believe, or at least have been absolutely certain until quite recently, that the developments in Western music since, let us say, the twelfth century—comprise a single more or less steady arch of progress.

I think it's fair to say that even scientists have had to learn that not everything that has happened in science since the Age of Enlightenment can be said to describe "progress." We musicians—though some would disagree with me—have, I think, an even greater problem in measuring, let alone claiming progress.

Indeed for me progress, in the sense that we usually understand that term, is either unmeasurable in music or irrelevant, or both. In my view we are left only with the knowledge (which, by the way, is quite sufficient for me) that each era, each epoch, each period, and beyond that different regions and countries and nationalities, have their own aesthetics and styles which may, in some final day of judgment, be deemed to be all equivalent—equivalent at least in their potential, or in their realization at the highest levels of genius and creativity.

But we have no means even of objectively verifying that statement or conjecture I've just proposed. For example, at first glance one might assume—and I'm sure most of us in this room would assume—that Beethoven and his aesthetics would rank some universal priority among musical forms of expression. After all, his music has moved human beings for 180 years and continues to do so unabated to this day, and presumably will do so in the foreseeable future. We, who belong more or less to that same culture that spawned Beethoven, like to think that his music contains some profound truth and communicative ability that transcends all people of whatever rank, class, or education or race. We ascribe to it "universality." And above all, we torment ourselves today with the notion that somehow Beethoven was on to something that we in the twentieth century (or for that matter musicians in the fifteenth century) cannot today and could not then achieve, right?

No, wrong! For even that widely held belief of Beethoven's universality is not altogether true. It is again conditioned by all kinds of inconclusive evidence and debatable assumptions. Upon closer inspection it turns out that, while *some* non-Western cultures, such as Japan's, have accepted Beethoven with open arms, just as they have obviously accepted Western technology and now excel in it, other cultures such as the Javanese or Indian or Arabic, cannot relate to Beethoven's music at all, even when all conditions for such acceptance appear to be propitious. All we can really say with certainty is that, *for the moment* and probably for some foreseeable amount of time, Beethoven's music seems to have a deep appeal, both potential and actual, for a wide segment of the population in *certain* cultures and *certain* human societies. But that's as far as one can go. One can neither prove nor disprove that Beethoven's popularity is universal, is permanent and invariant. It is, as the mathematicians would say, undecidable.

Music historians, musicologists and well-read musicians know that our history is filled with every kind of mistaken verdict, which in hindsight then often seems incomprehensible to us, be it an example of premature acclaim or belated recognition, of under-evaluation or overglorification, of acceptance in some places or by some people and not by other places and other people, and so on in endless variations. Thus we know—those of us who think about these things—that it is difficult even to assess the past. And therefore, how much more difficult—as I am being asked to do—to assess the present!

For even a cursory historical glance at our music-historical past reveals that not only have musical aesthetics defied unanimous interpretation or definition but, more crucial, the very essence and role of music itself has resisted complete, unequivocal defining, despite countless attempts for centuries (and indeed millennia) to do so. The most that can be claimed is that at *certain* times *certain* aspects of the total phenomenon of music have been *temporarily* defined. The earliest attempts go back to Alexandrian and Greek antiquity, to Ptolemy, Plato, Aristotle, and one Aristides Quintillianos, who all subscribed at least to the concept of dividing music into two equivalent complementary spheres, which they called *scientia* and *ars:* the science of music and the art or practice of music. This very broad and generalized initial definition was rearticulated in the Middle Ages by such music theoreticians as Saint Augustine, Guido of Arezzo, and Boethius, who amended the earlier concept of music to include the notion that two additional determinants were essential: those of *tone* (as an acoustical phenomenon) and *number* (as a mathematical element). Or to formulate it in another way, the two prerequisite factors by that definition, were *sensus* and *ratio:* the senses and the mind.

But it was in the final centuries of the Middle Ages, just before the Renaissance, that two further and, this time, opposing interpretations became current. It was the beginning of the split between the sacred and the secular, where the former sought to impose morality and truth as interpreted, of course, from a theological point of view, while the other would be satisfied with theoretical knowledge and practical ability. These restricted and essentially opposing viewpoints, though maintained in the Age of Humanism and the Renaissance, were nevertheless somewhat reinterpreted and broadened to emphasize the *practice* of music, particularly that of the vocal art as defined by ability and skill.

But it was not until the Age of Enlightenment that the ancient definitions and concepts of music finally gave way to an interpretation in which the *ratio,* the rational, the mind, embodied in man—as opposed to God and religion—and as delimited by *man's* intelligence, reigned supreme. And what is fascinating to note here is, on the one hand, once again the pre-eminence of mathematics in musical theory/practice, and on the other hand, a brand new ingredient at that time: the intelligence, the intellect of man, *Der Geist des Menschens* (as the Germans would have it), as particularly defined in the eighteenth-century formulation of man's mind as the *subject* and music as the *object.* And thus the Cartesian notion of "I think, therefore I am" came to the forefront in music.

Man was now perceived as standing in a central position in determining both the science and the art of music. This perception in turn allowed for yet another idea to surface, namely, the demand that music pursue the ideals of consonance and beauty. As Jean Jacques Rousseau put it in typically French fashion—this is a paraphrase translation: "Music is the art of combining sounds in a manner which is agreeable to the ear."

It is at this historical point—the last quarter of the eighteenth century—that a dramatic shift occurred, an almost 180-degree turn, as it were, ushering in the Age of Romanticism. Here *ratio* and the mind were supplanted by the expression of emotions and feelings as experienced by the individual, by the romantic hero personified by a Byron or a Beethoven. While such leading musicians as the composers Carl Maria von Weber and Richard Wagner and the famous German music historian Johann Forkel fully subscribed to that "romantic hero" point of view, they also tried to elevate music to the realm of the metaphysical and the transcendental. Of course, we all know that Wagner had his opponents, not so much in Brahms himself, but in Brahms's champion, the famous Viennese critic Hanslick, whose aesthetic was embodied in the motto "Music is nothing but sound-enlivened form" (in German, "tönend bewegte Form"), a reiteration of music as an absolute, abstract, *sui generis* expression. That split, exemplified in the music of Wagner and Brahms, was perpetuated through the first half of our twentieth century by those two latter-day giants of music, Schoenberg and Stravinsky.

Curiously Schoenberg, although caught up as a young man in the Wagnerian and post-Wagnerian fever, nevertheless was an equally ardent admirer of Brahms and his brand of classic formalism. One can say that, not only were the conflicts in Schoenberg's music between the aesthetics of Brahms and Wagner never fully resolved, but in a fascinating alchemical fusion and synthesis, Schoenberg managed to draw fruitfully upon both philosophies at various periods in his life. Still, when Schoenberg defined his aesthetic in words and not in music, he combined his basically romantic posture with one of the fundamental concepts of antiquity, namely music as an imitative art. And by that I don't mean mere program music, in other words an art not merely imitating nature's externals, but its *inner* expression through emotional and spiritual states.

Stravinsky, on the other hand, negated such a "romantic" viewpoint in favor of the old Middle Ages concept of music as pure expression of cosmic order. He said—and oddly enough his music often belies it—that music is incapable of expressing *anything*, even feeling, emotions, states of mind, let alone objects. Stravinsky felt that what we call the "expressive" in music, was but an illusion. Whether he maintained this belief even after he converted at the end of his life to Schoenberg's and Anton Webern's serial technique and its aesthetic is, oddly enough, not known. Stravinsky was to my knowledge never questioned on that subject and, as far as I know, he never voluntarily expressed himself on it in public.

Well, this brings us almost to the present in this brief capsule history of aesthetics and form in music. When I was a young composer in the 1940s, just beginning to discover myself aesthetically and artistically, it was made very clear to me and my generation that you absolutely *had* to choose between Schoenberg and Stravinsky as leaders, not only between the atonal serial technique of the one and the tonal neo-classicism of the other, but between their then very hotly debated philosophic and aesthetic positions. And I believe I was one of the first of my generation to say in response to that demand, "Nonsense! They are both great masters. I have much to learn from both."

Now, some forty years later, what seemed like such a complex and emotional choice pales by comparison with the present situation, where not two alternatives vie with each other, but dozens. Therefore it is particularly difficult right now to say anything definitive about the direction of music today, about form and aesthetics in this part of the twentieth century. Our century is now eighty years old and in these short eighty years, particularly since 1945, our century has witnessed an unprecedented proliferation and fragmentation of viewpoints, of musical aesthetics, of philosophies and techniques and schools of thought. Moreover, none has emerged in a central or leadership position. What we have instead is a broad spectrum of musical categories and concepts, all in a vigorous state of flux and cross-fertilization. This is an entirely historically unprecedented situation, awesome in its complexity, but aglow with the excitement of freedom—freedom of choice and of alternatives.

We are in my view on the threshold of some very exciting prospects. The certainties—or I have to say the *alleged* certainties of the avant-garde of but fifteen and twenty years ago—have given way to the relative uncertainties of the present state of eclectism and pluralism. We were so sure in the late 1950s and 1960s that the prevailing aesthetics and principles of form were not only superior but inviolate. Not that everyone subscribed to such dicta, as they were handed down by the avant-garde centers in Europe (and in our own country) and its would-be dictators. But in the professions, in the academies, and even the marketplace it was made pretty clear in which direction music was supposed to be going at that time. And the dominant aesthetic was that of absolutism, the belief in the unassailable priority of absolute music; and the dominant forms were those of what came to be called the "open form" and asymmetry. Pierre Boulez and Karlheinz Stockhausen were absolutely convinced that the "closed" forms of the past, those forms tied to the functional diatonicism of the previous two and a half centuries were exhausted, were dead, were obsolete; and the aesthetic ideals that went with those traditional forms, both the classic and romantic and the early twentieth-century models, were also dead and irrelevant.

Well, now at least, we seem to know better. It seems to me that we are and have been moving for about ten years toward a juncture where we are finally assessing in depth the innovations and experiments of the past seventy to eighty years, and both synthesizing and sorting out with an eye (and an ear!) toward discovering, *what* of all this recent past is of lasting value for now and for the future; and what on the other hand is unnecessary ballast. It seems to me that we are in a situation analogous to the one in the last third of the eighteenth century, when the previous innovations and experiments and various attempts to form a common musical language, a *lingua franca,* innovations which began with Monteverdi and culminated with Bach and three or four of his most gifted sons, resulted in a setting in which Haydn and Mozart had only to gather up various strands and pull them together into new forms, a new aesthetic, and a new musical language. Beethoven in turn was to profit immensely from their efforts. We have gone through a similar period in this century, of pushing the frontiers of musical techniques, of new forms and aesthetics—new ones it seemed every new generation, every new decade—to a point where we have exhausted our hunger for experimentation and are now hoping to reap the benefits of those hard-won skirmishes and battles.

As I have already implied in this brief retrospective of the history of aesthetics and music, not all of our achievements in recent decades can be described as gains. I for one am not sure at all that the espousal of the "open form," for example, first envisioned by Claude Debussy and the young Stravinsky and elevated to a veritable prescription and dictum by Boulez, constituted a gain over the closed forms of the past. It is more likely that the open form represents an interesting alternative or addition to our form arsenal and resources. The "open form," an exciting discovery early in the century, did away with recapitulation, with repetition, with the pre-eminence of a central unifying musical idea and, in short order, with the very concept and use of theme. Melody was soon to follow. In the first flush of excitement at what seemed like boundless vistas of freedom, i.e., freedom from previous form constraints, it was difficult to see that such a concept of highly individualized and totally open form had in it the seeds of its own eventual demise. For what now followed was the assumption by composers who regarded themselves as progressive, that henceforth *every* piece by *every* composer could have—and some said *should* have—its own indigenous form, unique and peculiar to itself.

It took some sixty years to come to the realization that that was an untenable idea. There are not tens and hundreds of thousands of new or different forms. One composer in a thousand is fortunate if he can in one lifetime invent *one* form which is special, which is unique, which is clear and strong, and which is worth remembering. And then there is the basic contradiction, that, if it is worth remembering and thus worth using again, then why persist in a theory that form must be forever self-renewing. It seems to me that such thinking—and this was only twenty and thirty years ago—represented a basic misunderstanding of the very meaning of form in music. In the old days form used to mean a structure, a *recognizable* structure, a basic mold into which a music could be poured, as it were. But form as viewed in the recent past in the concept of "open form," i.e., as something constantly variable, unpredictable, and finally anarchical, wasn't form at all. It was merely a container, a loose wrapping; and, alas, very often an indiscriminate *absence* of form.

It seems doubtful to me that certain traditional musical forms—whether the "closed" sonata or variation forms of classical music, or the Blues form and structure in jazz, or the "call and response" forms of many ethnic and folk musics around the world, particularly those of Black Africans and Native Americans—that such forms (and many more like them) will ever become useless or exhausted. Certain generic strains, even in music, seem to have survival built into their genes.

It is the task of the musical artist to know which these are. Ironically, many people, musical laymen, seem to already know. With the more recent recognition (in the last decade) that many of the traditional precepts of form can still be valid for our time, came the further recognition that we now have available to us a whole range of forms, running the gamut from the most sprawling and informal *wide open* forms—if we can manage to control them—to the most formalized, strict *closed* forms; and, of course, as usual a great many shadings and gradations in between. But I think it is also not precluded that, just as Josef Haydn took the embryonic symphonic form of his day and developed it into its first full flowering, so too today a Haydn of tomorrow may come up with a brilliant new form concept.

In the realm of aesthetics things are even more kaleidoscopic. We are looking today (in music) at a bewildering array—a giant rainbow—of aesthetic alternatives of every conceivable shade and coloring, from John Cage's aesthetics of "non-art," "indeterminate form," and "chance operations" applied to all elements of music, to the exquisite and subtle refinements of virtually total control of a Milton Babbitt. In his music, for the first time in centuries, form equals content and content equals form. With a thousand and one variations on and deviations from those two extremes of the aesthetic spectrum, we have thus a veritable plethora of alternatives to choose from. Again the dichotomy of apparent total freedom of choice and the awesome responsibility of how and what to choose.

So, the young composer of today doesn't have it easy. On the one hand he can do everything and anything, singly or in combination, as he wants. But on the other hand, what a frightening challenge that is. How much one must know to make an intelligent choice among so many alternatives in this veritable *cafeteria* of musical aesthetics and techniques. Add to this new media and instrumental techniques, such as computer and electronic music, and a whole new world, for example, of multiphonics on modern wind instruments—where you play two and three or four notes simultaneously—and you have an idea of both the richness and the inherent complexity (some would call it confusion or chaos) of the present situation.

But one very healthy sign is that composers are thinking about aesthetics and beauty again. Beauty was an inadmissible and bad word just twenty or thirty years ago. It was never discussed in a graduate class in composition. It was not long ago that the average young Turk, aided and abetted by the musical academicians that populate our schools and graduate music departments, equated aesthetics solely with technique. Such and such a technique begat such and such an aesthetic, and vice versa. It is true, of course, that certain techniques, if used in a very restricted way, will result in similarly restricted aesthetic frameworks; and that again is precisely one of the choices a composer can still make, if he wishes to do so. But in contrast to yesteryear, the erstwhile clear lines of demarcation between formal and aesthetic alternatives—which in the past were often hotly contested, as I mentioned earlier—have now become blurred and softened; and formerly inimical aesthetic viewpoints have begun to converge, to overlap, to cross-fertilize, or at the very least learned to coexist peacefully. Call that other approach "musical isolationism," "musical aryanism," or "maintenance of pure genetic musical strains": we had all that and more for years. But now, in my view at least, we seem to enjoy a more benign, open-vistaed outlook.

Of course, I don't wish to be trapped into talking only about what we call "classical contemporary music." For it is precisely the present-day reality, in which all kinds of musics coexist on the face of this globe and, I dare say, now even cohabit in new and wonderful pairings and combinings—maybe even "symmetry breakings"—which defines most accurately the present state of music, notwithstanding the fact that the entrenched musical establishments of our Western culture have not yet understood or accepted that reality. I regard the old notions of European-based "classical" music as being inherently superior to all other forms and aesthetics of music as totally untenable. I have a quite different view of music—I call it the "global view" of music—and in that global view there are only two kinds of music, as Gioacchino Rossini put it a century and a half ago: "good and bad." There is no inherent or automatic correlation between *categories* or *types* of music and quality or the lack of it. There is only a broad spectrum of musics and musical aesthetics ranging from the most ancient to the newest, from the popular, folk, ethnic, and vernacular to the serious, classical, symphonic—those are all hopeless misnomers—from the improvised to the fully notated. In all these different *kinds* of music there can be found the exalted creations of musical geniuses or the droppings of the mediocre and the feeble offerings of commercial parasites. No musical strain has an inherent claim on quality—or the absence of it. The genius of Beethoven and Bach or Louis Armstrong and Charlie Parker; the genius of those unknown musicians who eight or nine centuries ago created *Etenraku* and the other ceremonial masterpieces of Japanese *Gagaku;* or the genius of an African improvised drum ensemble, whose rhythmic complexity and virtuosity defy our Western capacities of musical comprehension, let alone emulation; or the genius of the eloquent nature-bound ceremonial music and dances of our own Pueblo Indians; or the heavenly serenity of Javanese Gamelan traditions—I could go on citing many, many more—all these forms of musical expressions are in my view and to my ears *qualitatively* equivalent.

That is why the motto of my own publishing company reads "All musics are created equal." By the way: not all musics *are* equal, but all musics are *created* equal, that is to say, *potentially* equal in quality. It all depends on who chooses to express himself through these kinds and categories of music and *how*. Our Western habit of regarding all those other musics and aesthetics out there as "primitive"—how dare we call African drum ensembles primitive!—or irrelevant is today a totally untenable and, in my view, embarrassingly ludicrous idea. How dare we sit in judgment, looking down from our European thrones on Javanese or African music or, for that matter, that of the Eskimos!

Well, one specific reason why we dare not came to us through the courtesy of science and technology and the ingenuity of modern man: namely, through the phonograph, the radio, television, satellite transmission and other twentieth-century breakthroughs in communication. Before our time we did have some sort of an excuse for our ignorance and prejudices about other musics. We literally could not know empirically what we can know today. Beethoven had no way of learning anything about the music of Tunisia or Afghanistan or of the Japanese court. Perhaps a Marco Polo or a Magellan had some inkling of the world's musical diversity, but neither of those gentlemen let on if they did.

But today's budding musician or lay listener is in the most fantastic position; he can go to the nearest record store and buy not one recorded example of such music, the music of Afghanistani mountain shepherds or whatever it might be, but literally dozens and hundreds of such recorded examples. No, it doesn't behoove us—nor cover us with much glory—to continue to ignore that whole other beautiful universe of music that lies out there beyond our own narrow musical horizons.

And here we artists have, I think, a lesson to learn from science, the history of science, and the aesthetics which motivate most creative scientists. Perhaps not in all the sciences and certainly not until recent centuries, but it seems to me from my limited knowledge that science and scientists accept much more readily than artists do the infinite diversity of nature. Certainly in biology, since Darwin, and more recently in physics, let's say in the work of a John Wheeler, diversity is understood as the very essence of life. It puzzles me why we Western musicians and artists cannot understand and cherish that same diversity in our field. Why can we not appreciate and teach that music, especially as a creative expression, is not merely an entertainment, a commercial commodity, a salable product, but in its highest forms an expression or a reflection of *life itself*, in all its myriad and infinite diversity? In my very amateur but nevertheless hopefully correct view of the history and development of science, I see an extraordinary parallel to what I have called a "global view of music": namely, if the development of the universe can be described as a never-ending breaking up—"symmetry breaking," as you call it—of cells and molecules and atoms and micro-organisms, then so too music has proliferated and fragmented in never-ending replication and differentiation, from that first moment when primitive man uttered his first primeval sound. That is a concept of music that I find beautiful, for it is consonant with the development not only of life but of humankind.

That seems to me to be a lesson which we musicians and artists in general have yet to learn from the scientists. The concept of "unity and diversity" is, alas, foreign to most musicians—at the very time that modern technology has given us the tools and the technical capabilities of appreciating just that.

There is much more to be said about other matters that I haven't even mentioned, although some of these we touched upon briefly in some of the discussions yesterday. For example, the function and utility of music; or the role of beauty in music, of simplicity and of complexity and the interaction of the two; and above all the deep mysteries of the creative process, about which we know next to nothing. I have described to you where matters stand now in respect to concepts, schools, styles, philosophies of music, and the extreme fragmentation that characterizes both the narrow field of Western classical music and the much larger field of global musical pluralism.

But there remains one other matter which I *must* mention, if only in passing, which is perhaps especially appropriate to discuss in the framework of this conference. It is a subject which I will have to state mostly in the form of questions, for it is not something to which I or anyone as yet has the answer. Moreover, it is a subject located at the very frontiers of both music and science and one which neither of us, musician and scientist, can study or solve independently. It is an area where we must collaborate and pool our talents and resources.

Formulated in the simplest terms the questions (or the problem) is: how precisely do we learn music? This may seem at first hearing a harmless or irrelevant or even stupid question. Many of you may say: of course, we know how we learn music. But the fact is we don't. The question's relevancy becomes agonizingly clear when we realize that there are many thousands of schools and tens of thousands of teachers who all teach music, without any of us really understanding *how* we learn and what the processes of acquiring musical skills, creative or re-creative, are. In other words, how can we teach something effectively if we don't know how and by what combination and accumulation of skills we learn what we learn?

I, for example, can tell you nothing about why I know what I know, and why I can do the particular musical things that I *can* do, and how I acquired those skills. And around this painful question cluster several other related ones. I cite only a few. What really are the formal dynamic principles that govern the creation of music—of this or any other period? What indeed is the *nature* of music? What is its essence, its content? What is the relationship between the human mind and human senses and the external world, which we presumably observe and, therefore, in partial ways understand and reflect? And why have we failed, thus far—as I pointed out earlier—to arrive at a complete formal and aesthetic theory of music? Ironically, the answers to these questions elude us intellectually, even though any musician will tell you that such intellectual abstractions constitute the most concrete experiential realities of music, realities which when we experience them convince us positively that a life in music is *life itself*, is beautiful, the fullness of living. In other words we *feel* these things, but we can't explain them; we can't give them an intellectually cogent, coherent definition.

I don't know whether it is entirely fair to ascribe one of our problems in music (and the other arts) to science: namely, the dichotomy between objectivity and subjectivity. I suspect that this issue has been with us a long time, since antiquity in fact, although not always articulated as a problem. But since the advent of modern science the conflict between objectivity and subjectivity, between the absolute and the relative (or the conditional), between intellectual knowledge and direct experiences, instinctual sensory responses—all these dichotomies have become priority concerns for the music theorist and aesthetician. As mentioned earlier, there has been no lack of attempts to resolve these questions and to unify objective and subjective values into a comprehensive and systematic whole. And yet, as I have also pointed out, our best minds have not yet achieved that goal of accurately defining the deeper formal and aesthetic realities of music. And because of this failure even the best of our music education, not only of professional musicians but of the general public, that is, our potential audience, is incomplete and inadequate unto the task. But how little we hear about these concerns and questions in the education offered by our music schools, conservatories and colleges and universities, virtually all careerist-oriented, where too frequently the acquisition of a minimum of utilitarian knowledge is considered sufficient to achieve a maximum of success and financial income!

I am sure you have noticed by now that my knowledge of science is hopelessly incomplete and inadequate. Nevertheless with my heart pounding, treading (I hope) not where only fools tread, I dare to suggest, in closing, an extraordinary and to me profoundly beautiful statement by Arthur Eddington. It reads: "All through the physical world runs that unknown content which must surely be the stuff of our consciousness. Here is a hint of aspects deep within the world of physics and yet unattainable by the methods of physics." Talk about the scientist as artist and philosopher and poet. Is this a hint, a possible postulate or hypothesis for further inquiry as to the true relationship of the objective, quantifiable world of *matter* and the subjective qualitative domain of the *mind?* For are not mind and matter ultimately derived from the same life sources? If I understand it correctly, quantum mechanics has shown that, at the basis of the physical universe, lies what—I hope I'm correct in this—quantum physicists have called the "ground state" or the "vacuum state"; and that this field and this conception is one not only of perfect order but of infinite energy and intelligence, whence all creation around us—so the theory suggests—has arisen. But there is another branch of science, psycho-physiology. If it is true, as the psycho-physiologists claim, that the thinking process of the human mind has at its basis what these scientists call the "state of least excitation of consciousness," then do we not have here a fascinating parallel between the physical universe and the human mind? That is, does not this suggest that these both derive from the same or single common source, ground source, *Urgrund,* as the Germans call it? And if *that* is the case, then do we not have here a postulate, a theory, for resolving those eternal conflicts that have plagued us forever and ever which I cited earlier: the dichotomies between objectivity and subjectivity, absolute and relative, physics and metaphysics, the external and inner worlds—indeed, science and art?

Perhaps this is what the greatest minds from Plato and Aristotle to Einstein, Eddington, and Wheeler have tried to get at or tell us: that the structure of the human mind and the external physical world are *at ground* identical or at least parallel. My question is: Can we musicians and scientists explore such an idea together, systematically and empirically, beginning perhaps with a conference such as this?

I am convinced we musicians need science's help. A more profound understanding of music depends on a more complete understanding of the workings and dynamics of the mind; and perhaps then, working at the very frontier of consciousness where mind and matter meet, we will begin to understand the creative act, what we call "inspiration" but have not been able to describe fully or define; understand how auditory perception, the acquisition of the skills related thereto, really functions.

How close we artists and you scientists can be, and at the same time how deep the symbiotic relationship between man and his environment can be, is exemplified in this oft-quoted and most touching statement and revelation of Einstein, an appropriate thought, I think, to leave with you as I conclude this talk: "The theory of relativity occurred to me by intuition, and music is the driving force behind this intuition. My parents had me study the violin from the time I was six. My new discovery is the result of musical perception."